HYDRODYNAMICS, SEDIMENT TRANSPORT AND LIGHT ...
CAPE BOLINAO, PHILIPPINES

Hydrodynamics, sediment transport and light extinction off Cape Bolinao, Philippines

DISSERTATION
Submitted in fulfilment of the requirements of
the Board of Deans of the Wageningen Agricultural University
and the Academic Board of the International Institute for Infrastructural,
Hydraulic and Environmental Engineering for the Degree of DOCTOR
to be defended in public
on Thursday, 19 June 1997 at 16:00 h in Wageningen

by
PAUL CADELINA RIVERA
born in Isabela, Philippines

CRC Press
Taylor & Francis Group
Boca Raton London New York

CRC Press is an imprint of the
Taylor & Francis Group, an **informa** business

This dissertation has been approved by the promotors:
Dr L. Lijklema, professor in Water Quality Management
Dr W. van Vierssen, professor in Aquatic Ecology

Published by
A.A. Balkema, P.O. Box 1675, 3000 BR Rotterdam, Netherlands
Fax: +31.10.4135947; E-mail: balkema@balkema.nl; Internet site: http://www.balkema.nl

A.A. Balkema Publishers, Old Post Road, Brookfield, VT 05036-9704, USA
Fax: 802.276.3837; E-mail: info@ashgate.com

ISBN 90 5410 408 2

Dedication

Itong along sa dagat,
tulad ng ating buhay,
kung minsan ay tahimik,
kung minsa'y magalaw.

Ganyan ang ating buhay,
kung merong kalungkutan,
paglipas ng hilahil,
merong kaligayahan.

Itong alon sa dagat,
itulak man ng hangin,
hindi makalampas,
sapagkat mabuhangin.

Bawat alon sa dagat…

Tinig na tumatawag!

- Fr. Hontiveros, S.J.

To my Parents and Brothers

Acknowledgment

The research work presented in this thesis is the result of a cooperation between Wageningen Agricultural University (WAU) and the International Institute for Infrastructural, Hydraulic and Environmental Engineering (IHE) both in the Netherlands, and the University of the Philippines Marine Science Institute (UPMSI) and Department of Meteorology and Oceanography (DMO) both in the Philippines. Written in the context of the project "Cooperation in Environmental Ecotechnology with Developing Countries (CEEDC)", this work was completed with the help of a number of persons to whom I am very grateful. Firstly, I thank Prof. dr. Lambertus Lijklema and Prof. dr. Wim Van Vierssen not only for initiating this project but also for their continued support and guidance throughout the course of my studies. Secondly, I am indebted to Ir. Gerard Blom for his untiring efforts to supervise the field research and to Dr. Erik de Ruijter for constantly coordinating every important activity within the project. I would like here to acknowledge also the good cooperation that Jaco Friedrich, Etienne LeJeune and Rob Ruiter showed while doing their practical period in the Philippines. Thirdly, I am grateful to Dr. Jorge de las Alas for motivating me in this important endeavor, to Dr. Mike Fortes for his persistent support especially during my stay at the UPMSI, to Dr. Josefina Argete whose kind efforts made the meteorological observations possible in Cape Bolinao, and to Dr. Mariano Estoque for helping us acquire some important equipment.

I am likewise grateful to a number of people in the Philippines especially at UPMSI for helping me, in one way or another, during the research. I am thankful to Prof. Edgardo Gomez, Drs. John and Liana McManus, Dr. Helen Yap, Dr. Gil Jacinto, Dr. Cesar Villanoy, Dr. Annette Juinio, Helen Capuli, Bert Estepa, Helen Bangi, Seb and Helen Dayao. In Bolinao, I am grateful to Elmer Dumaran, Tirso, Jack, Chris Diolazo, Bon Caasi, Boyet Elefante, Lagoy, Gene, Marlene and all the working people of the Bolinao Marine Laboratory. In the Netherlands, a number of people extended their kind support. At WAU, I am grateful to Drs. Jean Gardeniers, Marijke Kuipers, Leo Langelaan, Michiel Blind, Rob

Portielje, Morten Grum, Hans Aalderink, Theo Ywema, John Beijers, Ronald Gijlstra, Marjanne, Frits and Rasko. I also appreciate the kind assistance of Bea Jansen at the Dean's Office for International Students. At IHE, I would like to thank Dr. Peter Kelderman, Dr. Jan Vermaat, Dr. Michiel Hootsmans, Tony Mins, Dr. Henk Lubberding, Ineke Melis and Mr. Rien Schaakenraad. I am grateful as well to all the people in our student house (Daniel, Paul van der Weide, Dick, AnneMarie, Marije, Werner, and Lars) for their kindness during my stay in Wageningen.

Finally, I am grateful to my parents, brothers, sisters-in-law, and relatives in the Philippines whose constant encouragement and prayers helped me finish what I had started. To all of them, and to all the rest whom I have received support but whose names I failed to mention, a million thanks.

Abstract

Observational and numerical modelling studies of the hydrodynamics, sediment transport, and light extinction were undertaken in the marine environment around Cape Bolinao in the Lingayen Gulf (Northwest Philippines). Abundant with ecologically important seagrasses and benthic organisms, Cape Bolinao is presently threatened with siltation and eutrophication problems. For this reason intensive field measurements of relevant environmental variables which include currents, tides, temperature, salinity, total suspended solids (TSS), ash-free dry weight (AFDW), sedimentation flux, grain size distribution and organic content of bottom sediments, gilvin absorption, phytoplankton concentration, and light extinction were executed from August 1993 to June 1995. Laboratory experiments were simultaneously undertaken to determine the sedimentation and light extinction characteristics of various sediment fractions. Using time series and regression analyses, the results were analyzed and presented. A set of numerical models were developed and applied in the area around Cape Bolinao and the Lingayen Gulf. A prognostic model for the hydrodynamics, driven by realistic wind and tide forces, was developed independently for the cape (fine-resolution model) and the gulf (coarse resolution model). An operational open boundary condition based on the method of wave propagation is discussed. The hydrodynamical predictions were used, in conjunction with a diagnostic surface wave model, to force the suspended sediment transport model. The transport model, which is based on the time-dependent advection-diffusion equation, is third order accurate in space and time. For a realistic description of the suspended sediment transport process in Cape Bolinao, resuspension and sedimentation fluxes were included in the numerical model using existing parameterizations. The predicted suspended sediment concentrations were used in a diagnostic model for light extinction. This later model is based on the assumption that the contributions of the optically active components to the attenuation of the photosynthetically available radiation (PAR) are linearly additive. Calibration of the numerical models using field observations produced a set of parameter values which is deemed representative for the area of investigation. Using these parameter values, the overall model predictions were in good agreement with field observations. Finally, using the integrated model, the impact of river sediment loads (treated as a conservative tracer) in the Bolinao reef system was quantified.

Contents

Chapter 1

Introduction

There is at present a global marine environmental problem related to the discharge of sediments in coastal seas. The barely protected Philippine coastal zone is certainly threatened with a similar marine pollution problem. Naturally or anthropogenically triggered, the consequences of sedimentation can be enormously disastrous. The continuous cycle of resuspension and sedimentation, and the consequent reduced light penetration threatens the ecological functions of the biodiversity of the marine ecosystem. One example of such ecosytem in danger is Cape Bolinao situated at the mouth of the Lingayen Gulf (16°25 N latitude, 119°58 E longitude) in the northwestern coast of the Philippines (Figure 1.1). Cape Bolinao is characterized by a fringing coral reef area with interspersed islands and islets extending northward from the mainland province of western Pangasinan. It is connected to the South China Sea through the western part of the Luzon sea with no appreciable continental shelf due to sharp increase in the bathymetric contours. A very narrow shelf of several kilometer distance, the reef area around the cape is abundant with various underwater biota like seagrasses, seaweeds and corals which are presently endangered by siltation and the associated problems of nutrient transport and eutrophication. Siltation and eutrophication affect the marine biota in various ways. The major impacts include changes in the substrate composition, changes in nutrient concentrations and increased turbidity due to increased amount of suspended solids.

There is great need for knowledge on the physical environmental characteristics of the marine waters off Cape Bolinao, and generally of most Philippine coastal waters. However, they have not been studied in much detail. Information on the hydrodynamics and description of the marine physics necessary to study the Philippine coastal sea environment is scarce. The

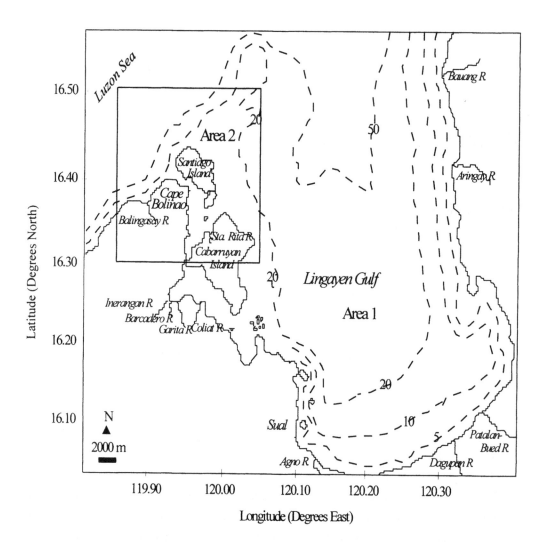

Figure 1.1. The Cape Bolinao marine environmental system geographically situated west of the mouth of the Lingayen Gulf. Bathymetric contours are given in fathoms.

reason for this is twofold. Firstly, physical oceanography and coastal engineering are relatively young sciences in the Philippines. Secondly, being unpopular sciences, very few scientists and engineers are devoted to conduct researches in these field of studies.

Contributions in the field of physical oceanographic researches in the Philippines started rather late and development is slow. One of the early researches on the physical oceanography of Western Philippine coastal waters started with the works of Wyrtki (1961) who investigated much of the South China Sea. Some observations which resulted to a qualitative and quantitative description of the dynamics of the area are provided in the NAGA report (1961). Additional information on the surface waters of the South China Sea is described in Levitus (1982). Monthly surface flows deduced from ship drift observations are presented in an atlas. Recently, observations are made on the dynamics of the Luzon Sea by the World Ocean Circulation Experiment (WOCE on-going). Aside from these few documented observations, there are some numerical modelling efforts which may provide some insight into the dynamics of the coastal waters of Northwestern Philippines. The baroclinic models of Pohlmann (1987), Shaw and Chao (1994) and Chao et al. (1994) present some quantitative descriptions of the water circulation within South China Sea and adjacent areas in the Philippines. Similar observational and modelling studies on the hydrodynamics of the Lingayen Gulf and of Cape Bolinao are presented in local literature. Villanoy (1988) deduced the circulation of the reef waters at Cape Bolinao from data observed on different occasions. On modelling, de las Alas (1986) conducted numerical simulations of the steady-state circulation in the Lingayen Gulf and described the wind-driven flow patterns of the gulf waters. Additionally, Balotro (1992) used a barotropic numerical model and presented the wind and tide-driven circulation patterns in the gulf. Information on the geochemical characteristics of the bottom sediments of Lingayen Gulf is provided in Santos et al. (1986). Heavy metal concentrations are measured in samples taken within Lingayen Gulf. Furthermore, using nuclear dating techniques, the sedimentation rate in the gulf was estimated. Chen (1993) studied the surface sediments of the South China Sea and provided some information on the geochemical and textural characteristics of sediments that exist in the Luzon Sea close to Cape Bolinao. The report provided some information on the sources of surface sediments at the South China Sea including the western Philippine sea. Concerning water quality, Maaliw et al. (1989) made a preliminary observational study on the general water quality characteristics of the Lingayen Gulf. In-situ field observations and laboratory measurements of relevant variables within the gulf are presented with some information on Cape Bolinao. Additional information on the marine environment of Cape Bolinao is provided by a number of biological and chemical research studies at the Marine Science Institute, University of the Philippines. However, observations on relevant physical processes are limited since they are treated as auxiliary variables.

The work presented here attempts for a detailed and in-depth study of the physical environmental characteristics of the marine waters at Cape Bolinao and at the Lingayen Gulf as a whole. Intensive field and laboratory investigations in the Bolinao area coupled with modelling studies on both the cape and the gulf as a whole are undertaken for this purpose. It should be noted that the present study is undertaken in conjunction with a simultaneous study on the ecology of the seagrasses at Cape Bolinao. The necessary environmental characterization of the area is provided by the present study. The objectives, relevance, scope and approach of this study is described in the following sections.

1.1.   Objectives and Relevance of the Study

This research work is itself a modest contribution to the physical oceanographic sciences of the Philippine coastal waters. In the present study, the marine environment of Cape Bolinao and the physical processes that exist therein are the central themes for investigation.

1.1.1. General and Specific Objectives

The general aim of this research study is to carry out observational and modelling studies to obtain a substantial data set for a detailed quantitative description of the existing physical environmental characteristics of the marine waters off Cape Bolinao. Relevant physical processes related to the hydrodynamics, suspended sediment transport and light extinction are described based on observational and modelling studies.

Furthermore, future conditions of the marine environment of Cape Bolinao are addressed by predictive models that are developed based on deterministic approaches. Such models are used for impact assessment studies of relevant marine environmental problems. The transport of suspended sediments and the associated underwater light extinction phenomena are particularly addressed in this work.

The specific objectives of this study are:

1.   To provide a quantitative description of the general circulation patterns, sediment transport processes and light extinction in the marine waters off Cape Bolinao.

2.   To develop a numerical model of the hydrodynamics, sediment transport and light extinction at Cape Bolinao and adjacent waters for environmental impact studies.

1.1.2. Relevance and Scope

The information provided in this research work is highly important in marine ecological studies. Relevant ecological research which needs this information includes studies on marine primary production, dispersal patterns of marine plankton (phytoplankton and zooplankton), growth of submerged vegetation such as seagrasses and seaweeds and growth of filter feeders and other benthic organisms.

The information from this study is also useful in assessing marine pollution in coastal environments. The distribution pattern of nutrients causing eutrophication problems in the coastal sea can be assessed using the results of the present research work. Furthermore, distribution patterns of heavy metals, including certain toxic compounds released into the marine environment can be described using the information provided in this study.

The scope of the study covers the following physical oceanographic processes observed at the coastal sea.

The hydrodynamics off Cape Bolinao and the whole Lingayen Gulf focuses on the long wave circulation induced by wind and tidal forcing. As the induced long-wave current is generally responsible for the advection and distribution of sediments in the coastal zone, detailed and in-depth study of the wind-induced and tide-generated currents and their interaction is presented. Additionally, effects of surface waves (windwaves) are addressed in the context of sediment transport processes.

The transport of sediments is mainly confined to the suspended load transport. Vertical transport via resuspension and sedimentation is discussed as much as horizontal transport (and re-distribution) due to the wind and tide-driven currents. Point sources of sediment loads from river discharges and their impact on the marine waters at Cape Bolinao are considered.

Light extinction is described in detail by taking into consideration most of the important factors affecting light absorption and scattering in the marine environment. These factors include suspended sediments (inorganic and organic), phytoplankton, gilvin and water itself. The organic fraction of the suspended sediments is taken into consideration by the observed ash-free dry weight (AFDW) concentration. Related characteristics of suspended sediments such as size spectra and fall velocities are considered to account for differences in the absorption and scattering of the underwater light field.

1.2.   Approach of the Study

Observational and modelling programs are designed to give the necessary information needed in this study.   Much of the effort in the execution of the observational and modelling programs is described below.

1.2.1. Observational Program

An intensive observational program was executed at many sites around Cape Bolinao during a period of more than 1½ years from late 1993 to mid-1995.   Selection of sites started in mid-1993.   These sites were chosen to cover the areas around Cape Bolinao which are mainly the areas where seagrass beds exist.   Different frequencies of observations are undertaken to cover the appropriate time scales of processes to be investigated.

Firstly, high frequency sampling of current speed and direction, tide, wind, water temperature, suspended sediment concentrations and light irradiance was executed to obtain information on the variability and dynamics of physical processes at time scales of several minutes to hours.   A mobile platform was installed at selected sites (Figure 1.2) and the necessary instruments which include a current meter with temperature sensor, automatic water sampler, pressure sensor, and light irradiance sensors attached to a datalogger were deployed.   The platform was moved to one of the four selected sites every 1½ months (or at least 1 month).   In this case, every site of deployment has observations for each of the two seasons (southwest and northeast monsoon season) prevailing in the area of study.

Secondly, weekly measurements at a number of sites (see Figure 1.2) of current speed and direction, water temperature, salinity, suspended solids concentrations, ash-free dry weight concentrations, gilvin absorption, phytoplankton concentration, sedimentation fluxes, and light irradiance were executed to obtain information on the dynamics of the investigated processes on a time scale of weeks or months.   These provide information on seasonal variations as well.   Aside from these variables, laboratory investigations on beam attenuation of water samples taken from the field were also done.   Furthermore, laboratory experiments on suspended sediment fractions according to fall velocity distributions were undertaken. The light attenuating characteristics of each sediment size fraction were then determined in the laboratory by spectrophotometric analyses of the corresponding beam attenuation of each fraction assumed.   Additional analysis was done for the contribution of algae to light attenuation.   Cultured samples of diatom species (*Isochrysis galbana* and *Chaetoceros gracilis*) were subjected to beam attenuation experiments as well.

Thirdly, quarterly investigation of bottom sediment samples taken from a number of sites at Cape Bolinao (see Figure 1.2) was undertaken. The size distribution of the sediment fractions assumed (e.g. sand, silt and clay) and their beam attenuation were determined in

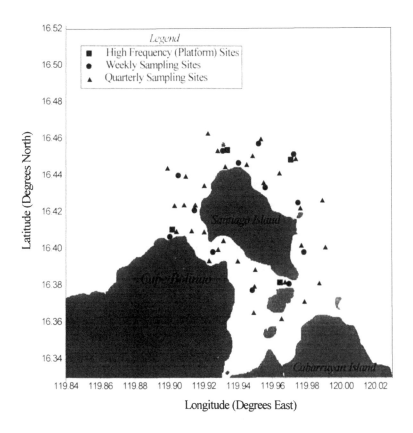

Figure 1.2. Map showing the study area and the measurement sites.

the laboratory. Aside from the above observations, precipitation was also observed and a total rainfall was recorded daily whenever there was rain at the study site. This last variable is important when considering processes related to sediment transport phenomena in the area.

1.2.2. Modelling Approach

The modelling approach takes into consideration the important physical processes and the developed model has the potential for operational purposes. As a rule, the numerical model must possess predictive capability without necessarily involving too much complexity. The choice of a modelling approach in this work is founded on such principles. After careful considerations of the preceding, a two-dimensional vertically integrated model for the hydrodynamics, sediment transport and associated light extinction is used to describe the dynamics of the concerned phenomena in the area of interest. Apparently, two areas have to be modelled, Area 1 covering Lingayen Gulf and Area 2 covering the waters off Cape Bolinao (Figure 1.1). The aim of this approach is to be able to describe satisfactorily the physical processes concerned at Cape Bolinao. While this is the main area of interest, the influence of the greater Lingayen Gulf to Cape Bolinao has to be taken into account.

To carry out such modelling, a coarse resolution model for Area 1 (Lingayen Gulf) with a grid distance of 1 km was designed. This is run with realistic forcing functions from observed wind and tide. A model for Area 2 (Cape Bolinao) with a resolution of 500 m is also designed independent of the Coarse Model. This fine model is also run independently with the same forcing functions. The influence of relevant processes (especially sediment transport) in the Lingayen Gulf to Cape Bolinao can be assessed using the coarse model. At the same time, localized processes at the main area of interest can be understood using the fine model.

Chapter 2

Environmental Profile of Cape Bolinao and Surrounding Areas

The environmental characteristics of Cape Bolinao and adjacent areas are typical for many coastal areas in the Philippines. There are however some obvious peculiarities in the general coastal environment of this particular area. The cape, being a small piece of land protruding into the open sea of the northwestern coast of the Philippines, has defined oceanographical characteristics not found in other coastal areas. The water quality, sedimentology and the coastal geometry are all affected by the dynamics of the waters around the area. There also exists a distinct topography surrounding the entire area which affects its meteorology and general marine environment. In particular, the monsoonal wind patterns especially during the northeast monsoon season are not similar to other areas in the Philippines due to the presence of mountains to the east and northeast. Furthermore, its geographical position has some bearing to the amount of precipitation it receives throughout the year, the air and water temperature variations, and the influence of weather disturbances. All these factors of various sources are manifested in the coastal environmental profile of the area. Description of the relevant environmental characteristics of the area of study and vicinity is provided in this chapter.

2.1. Coastal Geomorphology and Aquatic Environment

2.1.1. Topography

The bottom topography of Cape Bolinao can be seen partly from the bathymetric chart of Lingayen Gulf (see Figure 1.1) published by the Philippine Coast and Geodetic Survey. A finer resolution map of the bathymetry is shown in Figure (2.1). Depths of less than 30 m are generally observed around Santiago Island, the biggest island within the cape. North of the island is a vast coral reef area. A general feature of the reef flat is a relatively deep coastal lagoon near the shoreline decreasing in depth northwards until the reef crest and then increasing again in depth at the reef slope offshore. The reef crest is an important characteristic of the area as incident waves experience a range of conceivable phenomena

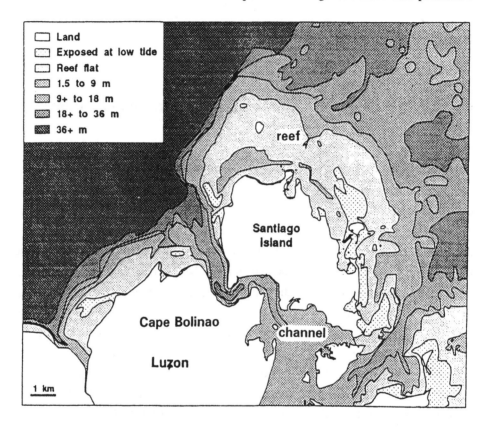

Figure 2.1. Bathymetric profile of Cape Bolinao marine waters (McManus et al., 1992).

including reflection, diffraction, refraction and occasional breaking, thereby sheltering the biotic communities within the reef flat from violent wave actions during stormy conditions. Another interesting feature of the bathymetry is a shallow portion extending about 15 km to the northeast of Santiago Island. This is described by McManus et al. (1992) as a subsurface barrier reef which is of hydrodynamic importance as well. A sharp increase in the depth distribution can be observed to the west and northwest of the area as shown by the bathymetric contours. From a shallow forereef slope depth of about 10 m, depths of more than 100 m can be observed already within 2 km offshore in those directions.

2.1.2. Morphology

The morphology of Cape Bolinao is complicated by the presence of islands and islets bounded by narrow channels. The coastal geometry of the cape is very irregular contributing to the complex hydrodynamical characteristics of the area. The biggest island of Santiago is separated from the mainland (Cape Bolinao) by a deep channel (see Figure 1.1). Depths there exceeds 20 m especially to the northwest. There is a however a general decrease with depth in the eastward direction within the channel presumably due to heavy siltation. Within the reef flat, there are small islands to the north and northeast of Santiago Island. Both islands are less than 1 km^2 in surface areas. To the south of Santiago Island is a relatively bigger island (area > 2 km^2) separated from Santiago by a narrow passage of less than 500 m. Together with several small islands to the south, they are separated from Bolinao and create a shallow semi-enclosed marine environment. Farther south, there is a narrow channel (width < 200 m and almost closed by an islet) separating mainland Bolinao and the biggest island in the Lingayen Gulf (Cabarruyan or Anda). Very narrow but deep, the small channel can serve as a connection between Cape Bolinao and the heavily silted southern coastal areas where several small rivers are located. This can be an important part of the coastal geomorphology of the area as it serves as a channel to Cape Bolinao for sediment transport by tidal currents. Aside from these system of islands and channels, there are several islets in the vicinity of the cape which may complicate the current and wave propagation patterns in the area.

2.1.3. Sedimentology

A description of the bottom sediments in the Lingayen Gulf give some insight into the sedimentology of Cape Bolinao. Santos et al.(1986) made a geochemical characterization of the bottom sediments at several stations within Lingayen Gulf. It was found that bottom sediments at the gulf have high metal concentrations due to mine tailings discharged by several mining companies of the Benguet Province east of the gulf and domestic inputs from coastal areas. Of particular significance is the increasing Cu content of the sediments from 4 m deep to the surface. Furthermore, through nuclear dating techniques, the sedimentation rate of the gulf was estimated to be in the order of 4.7 cm year^{-1}.

A recent study on the surficial sediments of the South China Sea was presented on the First Working Group Meeting on Marine Scientific Research in the South China Sea (1993) giving additional information on the sedimentology of the surrounding areas near Cape Bolinao. The percentage of clayey sediments was found to be generally higher than that of silt and sand. The result of the grain size analyses further indicated 40 % clay and 30% silt content of sediments gathered west of Cape Bolinao. The sediment mineral content in the area was also reported to contain quartz (15 %), plagioclase (20 %), illite (10 %), K-fledspar (5%) and amphibole (up to 15 %). The increasing amphibole content of the sediment towards the land was known to indicate the origin of the mineral from the main island of Luzon (North Philippines). Furthermore, the carbonate content of bottom sediments west of Cape Bolinao was reported to be generally less than 10 % (Chen 1993).

2.1.4. Water Quality

A water quality baseline study at the Lingayen Gulf (Maaliw et al., 1989) conducted in 1987-88 presents some insight into the state of the marine waters around Cape Bolinao. The observations made in several stations included measurements of primary water quality variables such as pH, temperature, salinity, dissolved oxygen, transparency, nutrients and heavy metals, BOD, as well as coliform and pesticide levels. Nutrients such as nitrite-nitrogen, nitrate-nitrogen, and total phosphorus were among the observed variables. Furthermore, heavy metals including lead, zinc, cadmium and mercury were analyzed in several stations. Figure (2.2) shows the locations of the sampling stations. Only the major rivers to the south of the gulf namely Agno River, Dagupan River and Patalan-Bued River were sampled for riverine and estuarine characterization while most of the stations are located near coastal areas bordering the gulf. Table II.1 shows mean values of several water quality variables measured in offshore, riverine and estuarine waters during the wet and dry seasons. Seasonal differences of water quality parameters are observed at several stations. Of particular significance is the finding that riverine and some estuarine stations showed consistently low D.O. levels for both dry and wet seasons. The low levels of D.O. observed

Variable	Offshore		Estuarine		Riverine	
	Wet	Dry	Wet	Dry	Wet	Dry
pH	8.1	8.0	8.0	7.9	8.0	7.9
Salinity (ppt)	29.6	33.4	14.5	29.7	12.8	19.3
Temperature (°C)	30.9	28.7	30.5	28.8	29.5	29.5
D.O. (mg/l)	5.27	4.46	4.20	5.09	4.14	4.00
Secchi Depth (m)	0.9	1.0	0.4	0.3	0.3	0.3

Table II.1. Summary of general water quality parameters during the wet and dry seasons (Maaliw et al., 1989).

apparently indicate an organic pollution where oxygen demand of oxidizable organic matter is high.

The mean nutrient levels in the water column measured in several stations within the coastal waters around Cape Bolinao are presented in Table II.2. Stations 11 - 13 are within the waters of Cape Bolinao (see Figure 2.2). The ranges in mean concentrations for the observed nutrients are very high. Maaliw et al. (1989) found a low average concentration of 0.18 μg/l to a very high mean concentration of 34 μg/l for nitrite. The authors also found mean concentrations which range from 0.92-26 μg/l for nitrate and 3.27-118 μg/l for phosphate. Elevated nutrient levels were observed particularly during the wet season. These relatively high nutrient levels were attributed to domestic effluents, organic and inorganic fertilizers from agricultural farms and fishponds and leaching from the soil (Maaliw et al., 1989).

Station	NO_2-N (μg/l)	NO_3-N (μg/l)	PO_4-PO_3 (μg/l)
11	5.39	9.32	6.63
12	8.22	6.59	18.34
13	6.04	4.05	5.30
14	1.94	2.71	13.37
15	2.39	2.87	15.63

Table II.2. Mean nutrient levels (data averaged from Maaliw et al. 1989).

2.1.5. Freshwater Influence

The major source of freshwater input to the whole Lingayen Gulf are from rivers (see Figure 2.2) which discharge considerable amounts of freshwater during the rainy season and with a constant lower supply during the dry season. There are five major rivers draining into the Lingayen Gulf. These include Agno, Dagupan, Patalan-Bued rivers in the south, and Aringay and Bauang rivers in the east. There are also small rivers (Inerangan, Garita, Barcadero and Coliat) in the west coast of the Lingayen Gulf which also contribute to the freshwater content of the gulf waters.

Of particular significance to the Bolinao area is the Balingasay river situated west of Cape Bolinao. The main river measures approximately 30 m in width and has an average depth of 3.5 m near the mouth. Measurement of water discharges during rainy days showed an average of 71.4 m^3 s^{-1} (high flow) and an average of 16.7 m^3 s^{-1} (low flow) during rainless days (EIA, 1994).

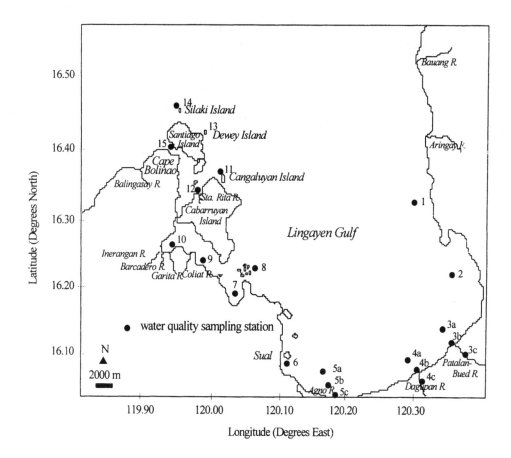

Figure 2.2.   Sampling locations of water quality baseline study at the Lingayen Gulf (Silvestre et al., 1989).

Another source of freshwater supply is groundwater. A qualitative groundwater resource map for the study area is shown in Figure (2.3). The amount of freshwater supply from groundwater is generally less than the total amount of river freshwater in the area. However, the almost constant supply of freshwater from underground sources is important in maintaining the salinity levels in the coastal waters. It should be noted that during the dry season when rivers do not discharge enough freshwater, the groundwater supply is responsible in preventing salt water intrusion that can cause adverse effects to near-shore biota.

Effects on the hydrodynamics of these freshwater inputs could be experienced during heavy discharges. In particular, the vertical stratification enhanced by density differences between freshwater and marine water could induce secondary circulations. In general however, mixing by the wind and tidal currents after a heavy discharge eliminates the formation of a permanent density-driven circulation.

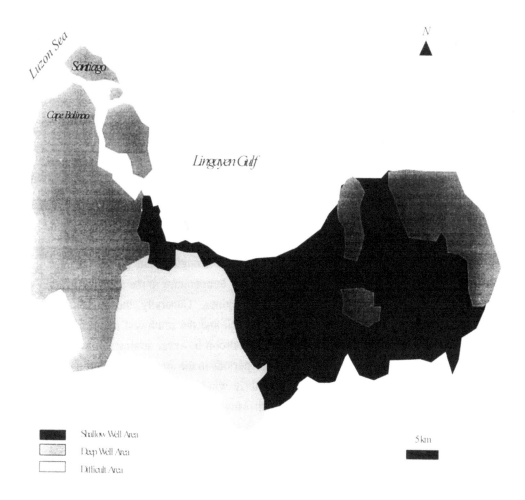

Figure 2.3. Groundwater resources map within the study area (EIA, 1994).

2.2. Meteorology and Oceanography

There exist some semi-permanent climatological patterns within the tropical regions which have important consequences on the environmental characteristics of the area of investigation.

The seasonal reversal of the winds during the southwest and northeast monsoon seasons for instance result in different circulation patterns in the surface waters off Cape Bolinao and the Lingayen Gulf as a whole. Furthermore, the passage of tropical cyclones in the area have significant influence on a range of coastal environmental processes. The general features of the coastal zone could, in fact, be the result of the prevailing meteorological and oceanographical conditions in the area. The generally great amount of precipitation due to meteorological disturbances in the area can also have significant impact on the general environmental characteristics of Lingayen Gulf. In particular, the sediments discharged into the coastal zone during heavy rainfall result in changes in the general water quality, bottom topography and water levels of the area. External pollution loads including silt, nutrients, heavy metals carried by rainwater upland have great impact on the coastal zone of the study area.

2.2.1. Meteorology

2.2.1.1. Monsoons and Sea Breeze Influence

The monsoons are basically reversing wind patterns brought about by the changes in the general atmospheric circulation and the sea surface temperature in the tropics. There exists two distinct features of the monsoons in the Philippines. Generally, the northeast monsoon persists during the months of November until April and the southwest monsoon during the months of May until October. The northeast monsoon however attains strength during the months of December until March and transition periods in the months of November and April are observed. Moderate to strong northwesterly winds during the peak of the northeast monsoon can be experienced in the area of investigation. The elevated mountain peaks of the Cordilleras to the north east of the study area seem to block the northeast winds resulting in an apparent change in its direction over the entire Lingayen Gulf area.

The southwest monsoon generally affects the area of investigation during the months of May until October but normally attains its peak during the months of June till September. The rainfall amount during this season is considerable. It is during this season that the passage of tropical cyclones are generally experienced and changes in the water circulations can occur due to a complex wind stress pattern as cyclones cross or influence the area of study.

Aside from the observed seasonal reversal of the winds due to the monsoons, observations at a meteorological station near the area showed the influence of the sea breeze. A semi-permanent mesoscale meteorological phenomenon known as the sea breeze results from the temperature difference between the sea and land. During daytime, when the land is warmed by solar heating, winds from the relatively colder sea rush onshore producing the sea breeze.

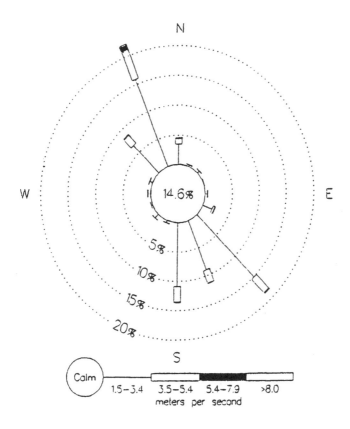

Figure 2.4. Annual windrose diagram representative of the study area (EIA, Bolinao Cement Plant).

Reversal of this atmospheric phenomenon occurs at night when the land cools down. The sea, with its relatively higher heat capacity maintains its temperature. This reversed temperature gradient between the land and the sea produce the land-breeze phenomenon with relatively weaker winds directed offshore. This land-sea breeze phenomenon prevails within the area as northwest-southeast reversing wind pattern. Figure (2.4) shows the windrose diagram (1961-1992) at a nearby meteorological station. A high percentage of occurrence (14.6 %) for 'calm' wind speeds not exceeding 1.5 m s^{-1} is observed. The highest percentage is recorded for wind speeds ranging from 1.5 - 3.4 m s^{-1} (> 15 %). The monsoon and sea-land breeze winds generally fall into these categories.

2.2.1.2. Tropical Weather Disturbances

The Philippines, being an archipelago bounded by the Pacific Ocean to the east and the South China Sea to the west, occasionally experience several weather disturbances developed in the two basins. These weather disturbances include the inter-tropical convergence zone (ITCZ), ordinary tropical depressions (low pressure areas) and severe disturbances like tropical cyclones.

Of particular significance are tropical cyclones which developed in the Western Pacific Basin during the southwest monsoon season. Often, low pressure areas develop and intensification to a tropical storm or typhoon occurs before crossing the Philippine islands. An average of 20 cyclones either directly hit or have significant influence on the atmospheric and coastal environments of the Philippines. The impact is mostly felt from strong winds and heavy precipitation. Considerable sediment transport is experienced by the area from eroded topsoils by surface runoff and river discharges. These can be seen as very turbid narrow areas near the coasts during heavy precipitation.

2.2.1.3. Air Temperature and Precipitation

Atmospheric temperature and precipitation are two meteorological variables which can have significant influence on coastal oceanography. Existing meteorological stations close to the area of study provide some information regarding these meteorological variables.

The atmospheric temperature at the study site has a mean value of 28°C and ranges from 18°C to 35°C (McManus and Chua, 1990). The minimum temperature is experienced during the month of January and the maximum during the month of April. There is a general increase in mean temperature from February to May and a decrease during the rainy months from June to October. Significant decrease in temperature is experienced in November until January.

Precipitation on the other hand is mostly experienced during the onset of the southwest monsoon season from May until October. The mean annual precipitation that the area receives is about 2500 mm. Highest precipitation occurs during the month of August, the peak of the southwest monsoon season with maximum rainfall of about 600 mm (McManus and Chua, 1990). A significant amount of precipitation is dumped when tropical cyclones cross the area in coincidence with the peak of the southwest monsoon.

2.2.2. Oceanography

2.2.2.1. Tides

The marine waters around Cape Bolinao have been reported by Wyrtki (1961) to have diurnal tides (Figure 2.5). Observations show however that the area generally have mixed tides

Figure 2.5. Geographical distribution of tides in the vicinity of the South China Sea (Wyrtki 1961).

with the diurnal components prevailing over the semi-diurnal components. This implies that there prevails one high water and one low water daily, but temporarily (once in a fortnight) two high waters and two low waters occur within a day which differ in height and high water time. Figure (2.6) shows a typical tidal observation at Cape Bolinao which confirm that the tide is a mixed one with prevailing diurnal components. The tidal range often exceeds 1 m at spring tide. Generally however, the area experiences a low tidal range of less than a meter especially during neap tides.

The tidal constituents responsible for the type of water level fluctuation experienced at Cape Bolinao are shown in Table II.3 with their corresponding periods in hours. Interactions between these four major tidal constituents and the topography of the area of study produce the characteristic curve shown in Figure (2.6).

Type	Tidal Constituent Name	T(Hrs)
M_2	semidiurnal principal lunar	12.42
S_2	semidiurnal principal solar	12.00
K_1	diurnal luni-solar	23.93
O_1	diurnal principal lunar	25.82

Table II.3. Prevailing tidal constituents at Cape Bolinao and vicinity (Wyrtki, 1961).

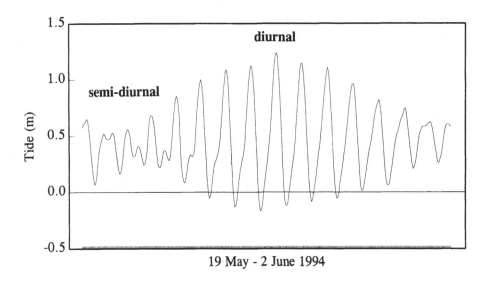

19 May - 2 June 1994

Figure 2.6. A typical tidal observation at Cape Bolinao (present study).

2.2.2.2. Water Temperature and Salinity

The surface water temperature in the area of study ranges from a minimum of 26 °C to a maximum of 32 °C with a mean value of 29 °C. Normally, the general trend in the water temperature is slightly out of phase with the trend in atmospheric temperature. McManus et al. (1992) reported a periodic pattern of temperature with maximum values in June and July just after the summer and minimum values in January and February during the winter. Surface temperatures can become temporarily low however during the passage of weather disturbances in the months of June till October. Temperature gradient in the vertical which leads to stratification of the water column is experienced particularly in the deeper portions of the study area but not in the shallow reef waters at Cape Bolinao. These temperature differences from the surface to the bottom which may have some influence on the hydrodynamics of the marine waters in the study area are brought about by strong solar heating at the surface during the dry season and cooling during heavy precipitation. The later source of temperature gradient may lead to instability in the water column. However, mixing due to the wind breeze and monsoonal winds tends to stabilize the water column and slight temperature gradients that may still occur do not contribute significantly to the vertical stratification and consequently to the instability of the water column.

The salinity diminishes with increasing freshwater input and ranges from a minimum of 26.8 ppt during the wet season to a maximum of 34.6 ppt during the dry season (McManus and Chua, 1990). Heavy rainfall with strong freshwater discharges from rivers contribute to the low salinity values. On the other hand, high salinity values during the dry season are experienced due to minimum freshwater discharges and enhanced evaporation.

Chapter 3

Monitoring the Marine Environment off Cape Bolinao

Monitoring the marine waters around the area of study is an essential part of this research. The physical processes at the coastal zone, localized and very variable as they are, can only be understood more clearly from an extensive field and laboratory observational program. Quantification of variables related to hydrodynamics, sediment transport and light extinction processes can be obtained from such observations. These measured variables can be very useful in understanding various physical processes of interest. At the same time, the modelling studies which are another important part of this research, rely heavily on such measurements. Model calibration and sensitivity analyses, in order to arrive at sensible results, can only be carried out with the aid of accurately determined quantities. The characteristic time and spatial scales of relevant processes are key factors in the monitoring program. It is likewise important that observations be carried out long enough to reveal variability of quantities or processes on longer time scales.

The execution of the observational study both in the field and in the laboratory is described in this chapter. Description of the general methods used are provided. Detailed procedures are available in Appendix 1 and some methods which are commonly used are given due citations.

3.1. Hydrodynamics and General Oceanography

The hydrodynamics of the waters off Cape Bolinao is governed by the interacting wind stress and tidal forcing. Motion due to density differences from temperature and salinity variations are of secondary importance but nevertheless observations are made to establish a complete data set on the variables responsible for the complex water motion in Cape Bolinao and surrounding areas.

3.1.1. Currents

Current measurements were done using the Niskin Winged Current Meter (NWCM Mk II) (General Oceanics, Inc.). It is a battery-powered self (RAM) recording current meter which measures current by the angle of tilt of its own housing. The wings of the housing orient the meter with the current and the direction is determined by a solid-state flux-gate compass. It uses a vector-averaging method where the internal microprocessor computes the average east and north components of current from a number of individual readings and records only those averages in the RAM (NWCM Operating Manual, 1991). The meter incorporates a real-time clock and uses the Universal Coordinated Time (GMT). When measurements are taken, an average is recorded with the GMT time (8 hours behind the Local Time).

The speed sensor is a force-balance tilt sensor. With the standard fin of the housing, the accuracy and resolution of this sensor are both equal to $\pm$ 1 cm/s. The current direction is measured with a three-axis flux-gate compass and has an accuracy of $\pm$ 2 degrees and a resolution of $\pm$ 1 degree. The meter was calibrated by the manufacturer before deployment.

For a coherent data set, current measurements were done together with other necessary environmental variables (e.g. light, water level, water sampling) in a platform. This was done for most of the measurements except in deep areas where current measurement was necessary and hence done independent of other meters. In this case, mooring is done with an anchor and a float (see Figure 3.1).

It should be noted that current measurements in shallow waters are prone to noises. A significant part of this noise comes from wave generated orbital movements whose magnitudes are stronger near the surface. When wind is strong, the wave-induced orbital velocities can be high and the actual horizontal current measured at the level of the current meter is not the real long-wave current of interest. Aside from this, the motion of the mooring line and the float may introduce noises which could mask the actual wind and tide generated horizontal currents. To overcome such noises, the meter has to be installed in deeper areas where wave effects are minimal. In shallow areas where most measurements were done, the meter is fixed to the platform avoiding the effects of the motion of the mooring line and float.

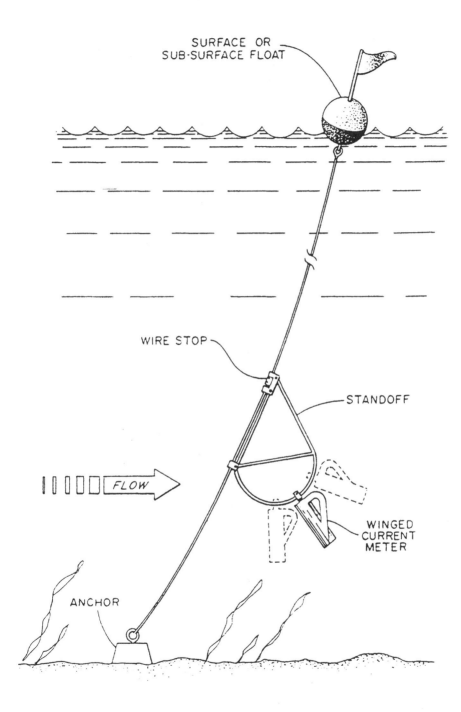

Figure 3.1. The self-recording Niskin Winged-Current Meter (NWCM) in a typical deployment in the coastal sea (General Oceanics NWCM Manual, 1993).

3.1.1.1. Tidal Currents

The tides induce a periodic horizontal water motion reversing in direction with periods corresponding to the periods of the dominant partial tide causing the motion. The NWCM current meter measures the magnitude of such periodic motions with its tilt sensor. The direction is then measured with its compass where a standard fin is oriented with the prevailing current direction.

This periodic component of the current is easily recognized from the data measurements when wind is weak. As wind measurement is done simultaneously, it is possible to determine the tide generated components of the flow in some occasions.

3.1.1.2. Wind-Induced Currents

Wind stress creates a horizontal gradient in the water surface. The drift currents due to the wind is random in nature due to the randomness of the wind stress causing it. The NWCM measures the total current and often in shallow water, the records would show high magnitudes of wind-driven current. These can be seen as elevated currents superimposed on the periodic component generated by the tide.

It should be noted that the current fluctuations due to short period windwaves are smoothed out by allowing the NWCM to take burst samples. Before deployment, the NWCM is set with a burst interval of 2 seconds and 8 samples per burst. Then a 5-minute average of the samples is recorded. In this way, noises introduced by the short waves are minimized.

3.1.2. Tides and Water Level Variation

Variations in water level is predominantly caused by tides which are basically long gravity wave with periods of several hours. Small variations are however caused by short period waves from the wind. These can be seen as small fluctuations with periods of some seconds superimposed on a periodic tide record. Measurements of water level were done with the WIKA Druckmeßumformer pressure sensor (Alexander Wiegand GmbH & Co., F.R.G.). The pressure signal from the water is measured by this sensor in milliamperes (mA). The sensor can measure signals from 4 - 20 mA. The signal is then recorded in a datalogger (CR10 Campbell Scientific, England). It should be noted that the CR10 datalogger records data in millivolts and the wires of the pressure sensor has to be fitted with appropriate resistors to transform signals from milliamperes to millivolts. The CR10 is then programmed to convert the signals into water levels in cm. A 1-millivolt signal corresponds to 1 cm water level. The maximum water level that can be measured by the sensor is about 250 cm which in most cases is capable of measuring the tide (maximum range is about 1.3 m in Bolinao) and windwaves. The windwaves may reach a 1 m range in the shallow waters of Bolinao during strong wind events.

It should be noted that the sensor has an accuracy of about 1/10th of a centimeter. This would make it ideal even for small wind generated waves. However, the high frequency needed to obtain a reasonable measurement of windwaves limits the use of the datalogger for wave measurement alone. Since there are other sensors attached to the datalogger and simultaneous measurement is important, the frequency of pressure sampling is reduced. Average values of 2-minute sampling interval were recorded and data analyses would reveal only the water level due to the tide. Measurement of wind-generated waves was not successful at Cape Bolinao.

3.1.3. Temperature

The NWCM is equipped with a temperature sensor and continuous measurement is done simultaneously with current speed and direction measurement. The sensor is an aged linear thermistor (Yellow Spring Instrument Co. Type 44202) and has an accuracy of $\pm$ 0.25 °C and a resolution of $\pm$ 1/64 °C. The range of measurable temperature is from - 5 to + 45 °C. The maximum temperature in Bolinao does not exceed 35 °C and the sensor is most suitable for most of the continuous temperature measurements done. Similar to current, 8 samples per burst with a 2-second burst interval were measured and 2-minute average values are automatically calculated and stored.

For non-continuous (weekly) measurements of temperature, a digital dissolved oxygen (DO) meter equipped with a temperature sensor is used. The temperature sensor of the DO meter (Yellow Springs Instrument Co., Inc.) has an accuracy of $\pm$ 0.2 °C. The measurable temperature is -4 to 45 °C which is well within the range of the water temperatures at Cape Bolinao.

3.1.4. Salinity

Most of the measurements of salinity were done with the same DO meter used in the temperature measurement (Yellow Spring Instrument Co., Inc.). A salinity sensor is fixed with the digital meter and measurement is done simultaneously with temperature. The resolution for salinity is 0.1 ppt and the accuracy is about ± 0.1 ppt. The range of measurable salinity by the sensor is from 0 to 40 ppt which is generally well within the range of observed salinity at Bolinao.

During the last months of the field work (January - June 1995), a different method to measure salinity was used. Water samples which are taken from the field were brought to the laboratory. The samples were properly covered to ensure that no evaporation occurs during the time that elapsed before measurement. A hand refractometer (ARGENT Chemical Laboratories) was then used to measure the salinity. It has a lesser resolution of 0.5 ppt and an accuracy of ± 0.5 ppt. In general, however, the resolution of this type of salinity meter is acceptable for obtaining long term salinity trends.

3.2.   Meteorology

Part of the measurements done in this study include meteorological variables. These are needed to obtain information on the relative strength of wind and variability in direction and rate of precipitation. The wind speed and direction are important driving forces for wind-induced currents and sediment resuspension. On the other hand, knowledge of the rate of precipitation in the area is important when considering surface land erosion and riverine discharges in sediment transport studies. The following sub-sections refer to such measurements of meteorological variables deemed necessary for this study.

3.2.1.   Wind Speed and Direction

Wind speed was measured using the speed sensor Model TV-114 (A.C. Generator) of Texas Electronics, Inc. USA. This is basically a three-cup anemometer constructed of gold anodized aluminum to prevent corrosion thereby making it suitable for coastal measurements of wind. Long term maintenance-free operation is characteristic of this sensor. Furthermore, a precision alternating current generator is employed which completely eliminates contact between rotating and stationary generator components (Texas Electronics, Inc., Wind Speed Sensors Manual). The signal from this sensor can be recorded using a datalogger which can convert the A.C. to D.C. signal. The Campbell CR10 datalogger was used in recording the signal from the sensor. The CR10 was programmed to convert the D.C. signal into wind speed in m/s with the appropriate multiplier and offset (constant) included. These were obtained when the sensor was calibrated with the ES-020 wind sensor of the Department of Meteorology and Oceanography of the University of the Philippines before deployment.

It should be noted that the sensor has a threshold nominal response of 2.0 mph. This means that wind speed below this is not properly measured. However, wind speeds of over 100 mph is measurable and the sensor can be used during typhoon occurrences which is characteristic of the area during the southwest monsoon season. The accuracy of the sensor is about $\pm$ 2 % with temperature limits of -50 °C to 50 °C and humidity limits of 0 to 100 %.

For wind direction, the Model TD 104D of Texas Electronics was used. The sensor utilizes a long life hybrid potentiometer mechanically coupled to the wind vane shaft. A very narrow 3° gap exists between the ends of the potentiometer (Texas Electronics, Inc.). The D.C. signal from the sensor is also tapped using the CR10 datalogger. Conversion of the D.C. signal into appropriate direction signals was made possible with the programmable CR10 datalogger. The operating range of the sensor is from 0 to 360° with a 3° dead band between the ends of the potentiometer. The nominal response of the sensor in terms of starting accuracy is $3.6 \pm .39°$ with the same operational specifications as the speed sensor.

3.2.2. Precipitation

The amount of precipitation was measured using a graduated cylinder with a diameter of about 2.5 inches. During a rain event, the volume of rainwater collected by the cylinder was recorded. This volume of rainwater was then converted into precipitation units in mm taking into consideration the effective surface area of the cylinder. There was no record of rainfall duration by any means and the rate of precipitation can only be estimated using daily average rainfall rates (in mm/day).

3.3. Suspended and Bottom Sediment Characterization

The variables relevant for the sediment transport part of this study includes the range of total suspended matter, either organic or inorganic, living or dead materials.

In addition to these variables, the filter-passing dissolved organic substance generally known as gilvin is also determined for subsequent light extinction studies.

3.3.1. TSS and AFDW Concentrations

For continuous measurement of TSS and AFDW, water samples were taken every 4 hours by an automatic water sampling device (ISCO Model 3700). The ISCO model 3700 is a programmable sampler powered by a rechargeable 12 VDC NiCD battery. It allows sequential collection of 24 samples with volumes of 1 liter each. A platform was installed to accommodate the sampler and with a 3 m hose, sampling of water at mid-depth was possible. It has to be noted that deployment sites in Bolinao (reef areas) are shallow and the low tide level should be taken into consideration so that the sampler hose is not exposed.

The water samples collected by the ISCO sampler were retrieved twice a week (Monday and Thursday). Analyses of TSS for the 1 liter samples in the laboratory follows the Dutch standard (NEN-methods).

For the weekly sampling, water samples are taken manually using 1 gallon plastic containers (available in many shops). These grab samples are analyzed for both TSS and AFDW including Gilvin, Chlorophyll-a, and beam attenuation. Analyses for TSS and AFDW follow the same NEN method.

It should be noted that for AFDW analyses, the dried samples (measured for total dry weight in Bolinao) are brought to Manila (Marine Science Institute) for ignition since there was no available furnace at the site during the time of sampling. Care is taken that the dried samples are not contaminated by dust. Properly sealed petri-dishes are used for transporting the dried samples.

3.3.2. Phytoplankton Concentration

Weekly, the phytoplankton concentration was determined by measuring the chlorophyll-a content of 2-liter water samples taken from several sites around Cape Bolinao. Initial measurements indicated a very low amount of chlorophyll-a from 1-liter samples. The concentrations obtained from small volumes are sometimes below the limit of detection. In succeeding measurements, 2-liter samples were used to obtain higher chlorophyll-a concentrations.

The procedure used for measuring chlorophyll-a is based on the Manual of Chemical and Biological Methods for Seawater Analysis (Parsons et al. 1984). The method, with slight modification is outlined in the appendix. The spectrophotometer used for this purpose was the Spectronic 20 (Milton Roy Company). The model (analogue type) is a single beam spectrophotometer with an overall wavelength range of 340 nm to 950 nm, covering most of the desired ranges for estimating the chlorophyll-a concentration.

3.3.3. Sedimentation Flux

In situ measurement of sedimentation flux was done by installing sediment traps at the designated sites at Cape Bolinao. The traps have a diameter of 5 cm and a length of 30 cm. This corresponds to a diameter/length ratio of 1/6 which is generally acceptable to avoid resuspension of collected matter in turbulent areas. The traps were deployed in the designated sites and were emptied each week. The total dry weight of the collected particles was determined and the sedimentation flux was estimated by dividing the total dry weight by the surface area of the traps and the period of installation.

Installation of traps in shallow areas are done by fixing them to wooden sticks which are hammered to the sediments. As a rule, the opening of the traps should not be exposed at low tide in shallow areas. On the other hand, deployment of traps in deep areas was done using lines. The trap was fixed at a certain level (mid-depth), a rock served as anchor and styrofoam as floats.

3.3.4. Grain Size Distribution and Organic Content

Quarterly (every 3 months), sediment samples from several sites around the area of study are collected for analyses of grain size and organic content. Surficial sediments (depth of which do not exceed 10 cm) are obtained using a PVC tube a diameter of 5 cm. These were dried in the oven at a temperature of 60 °C for over 24 hours (period depends on whether the sediments are dry enough for sieving). These were dry sieved with sieves of mesh sizes of > 3.35 mm, 600 μm, 250 μm, 125 μm, 63 μm and < 63 μm.

Of these dried samples, a small amount (10-20 g) was stored in carefully sealed petri-dishes and transported to Manila for measurement of the organic matter content. Before these samples were put in the furnace, the dry weight was taken. Then they were dried at 550 °C for a period of 6 hours. The organic content was then determined by the ash free dry weight of the sediment samples.

3.3.5. Bottom Sediment Fractionation

From the bottom sediment samples taken each quarter of a year, an attempt was made to determine the relative content of sand, silt and clay for each of the designated sites around Cape Bolinao. The whole procedure is outlined in Appendix 1.

3.3.6. Suspended Sediment Fluxes and Composition

For the purpose of determining fluxes and composition of suspended matter, sediment traps (4 to 5) were installed at mid water depth taking into consideration the lowest water level during ebb so that traps are not exposed. The sediment traps were emptied each week and the contents analyzed in the laboratory.

In the procedure for determining sediment fluxes, three sediment fractions were considered namely sand, silt and clay. These sediment fractions were considered by taking appropriate ranges of fall velocities in literature. The general procedure is shown in the appendix.

Average fluxes were determined by taking into consideration the volume of sub-samples, total volume of sample, period of installation, and effective surface openings of sediment traps. For chlorophyll-a, the same procedure as in Section 3.3.2. was used and approximate algal fluxes were also determined considering the above parameters. Gilvin absorption of the filtrate is a separate measurement in the interest of light extinction investigation (see 3.4.3).

3.4.   Underwater Light Conditions

The underwater light conditions at Cape Bolinao were characterized by in-situ measurements of irradiance and laboratory investigations using existing spectrophotometric methods. Being affected by a number of factors, the light extinction phenomena at the area of study were studied by taking into consideration the independent characteristics of these factors. Laboratory investigations were then necessary for such undertaking as these can not be done by in-situ field measurements.

Generally, the Photosynthetically Active Radiation (PAR) is of particular importance in environmental studies hence light measurements were executed to cover this range of the light spectrum (400-700 nm).

3.4.1. Downward Irradiance

The underwater irradiance was measured with two types of sensors. At the platform where continuous measurement is necessary, the Bottemanne sensors were used (Bottemanne Weather Instruments, Amsterdam). These are basically photosynthetic radiation meters with a quantum response calibrated within the visible range of the light spectrum. Radiation units given by this PAR meter are in photons, expressed as μmol s^{-1} m^{-2} (1 μmol = 6.022 x 10^{17} photons). The signal from this sensor given in mV is recorded by the CR10 datalogger at the platform (1 mV = 10 μmol s^{-1} m^{-2}). An appropriate program for the CR10 was made to obtain measurements with time intervals of 2 min. The accuracy of this type of sensor is $\pm$ 2% and it can measure irradiance from 1 to 2000 μmol s^{-1} m^{-2} which is generally acceptable in relatively less turbid marine waters.

Manual measurements of irradiance (weekly routine) were done with the LICOR sensor (LICOR, USA). This is basically a cosine-corrected irradiance sensor similar to the Bottemanne which covers the PAR range as well. A handheld datalogger from the same company is used to record instantaneous measurements. With an appropriate calibration constant, the voltage signal from the sensor is converted into irradiance units of μmol s^{-1} m^{-2}.

3.4.2. The Vertical Attenuation Coefficient

The vertical attenuation coefficient (k_d) is known to be the best single parameter by means of which different water bodies may be characterized in terms of the available photosynthetically useful radiant energy. Basically, two values of irradiance are measured in order to get an estimate of the vertical attenuation coefficient. This necessitates measurements at two levels in the water column, one near the surface and another below the first level noting the difference in depth. The quantification of k_d follows from the general law of Beer, i.e. $k_d = 1/z \, (\ln E_o - \ln E_z)$ where E_o and E_z are respectively the observed irradiance values near the surface and at depth z (below the surface).

3.4.3. Gilvin Absorption Coefficient

An important contribution to light extinction in coastal waters comes from gilvin. Gilvin is basically filter-passing dissolved organic matter which contribute to the yellow color of many natural waters. Terms such as yellow substance, gelbstoff, humic or fulvic acids are common names for this substance. It strongly absorbs light below the infrared range.

Gilvin absorption was determined in the laboratory from filtered water samples taken from the field. Water samples were filtered using the Whatman GF/F filters and a small amount of the filtrate (50 ml) was stored in tubes for subsequent measurements in a spectrophotometer (Spectronic 20, Milton Roy Company). The absorbance at the 380 nm wavelength generally gives a good indication of the maximum concentration of gilvin and this was scanned for filtered samples at the spectrophotometer Spectronic 20. The absorption coefficient can then be estimated using the standard spectrophotometric determination taking into consideration the actual cuvette size or pathlength used.

3.4.4.   Specific Beam Attenuation Coefficients

Specific beam attenuation coefficients of suspended sediments from the fractionation experiment, water samples from the field, and cultured algae were also determined in the laboratory. Three different kinds of suspension from the fractionation experiment were subjected for the analyses, i.e. sand, silt and clay. The following sub-sections describe the procedure for each of these samples.

3.4.4.1. Suspended Sediment Fractions

From the result of the fractionation experiments (see section 3.3.5), small amounts (about 50 ml) of the sand, silt and clay samples were stored. The absorbances of these samples were scanned using the Spectronic 20 and the beam attenuation coefficient calculated using same equation for gilvin (see 3.4.3). The specific beam attenuation coefficients were then determined by dividing the beam attenuation coefficient by the observed dry weights of each of the sediment fractions. As a small scanning step is not very practical with the available spectrophotometer, measurements at 50 nm interval were deemed acceptable from the 350-700 nm range.

3.4.4.2. Water Samples

Water samples were taken from the field and weekly determination of the specific beam attenuation coefficients was undertaken. The procedure is meant specifically to obtain information on the relation between observed light extinction coefficients and the specific beam attenuation coefficients. Water samples off Cape Bolinao are generally clear for beam attenuation measurements and only four turbid sites were sampled, namely the sites close to the mainland of Bolinao. The procedure for the determination of the specific beam attenuation coefficients follows that of section 3.4.4.1.

3.4.4.3. Diatoms

The independent contribution of phytoplankton to light extinction in the coastal waters off Cape Bolinao was also determined. Cultured samples of diatoms (*Isochrysis galbana* and *Chaetoceros gracilis*) at the Bolinao Marine Laboratory were subjected for this purpose. These phytoplankton species were used because they are known to be the dominant species of the algal population in Bolinao. Samples of 500 ml each are taken from 1 to 2 weeks old cultured samples. 450 ml was filtered using Whatman GF/C filters and their concentration determined following section 3.3.2. The 50 ml was subjected to spectrophotometric absorbance analyses, scanning the wavelength from 350 to 600 nm with a scan step of 50 nm. The Spectronic 20 (Milton Roy Company) used for this purpose is of limited capability.

The specific beam attenuation coefficients were then determined from the observed beam attenuation coefficients divided by the observed concentrations of each of the algal species.

Chapter 4

Hydrodynamical Modelling of Flows and Waves in the Coastal Zone

The hydrodynamical modelling study described in this chapter covers the long-wave induced circulation phenomena at the Lingayen Gulf and at Cape Bolinao with some relevant information on surface waves as necessary for the processes discussed in the following chapter. The coastal circulation induced by the wind and the tide is discussed independently. However, the actual hydrodynamical modelling done in this study treat both effects simultaneously in order to assess the flow patterns in the areas of interest subject to realistic driving forces. The chapter begins with some theoretical foundations followed by model formulation. The modelling study provides important descriptions on the treatment of open boundaries which are essential in dealing with open coastal environmental systems such as the ones considered here.

Two cases of hydrodynamic models are presented. The whole Lingayen Gulf is modelled using a coarse resolution grid and hence called Coarse Resolution Model. This model encompasses Cape Bolinao but the horizontal resolution is not fine enough to fully understand the dynamics of a small-scale area such as Cape Bolinao whose coastal geometry is complicated. This is the reason why a higher resolution model is also developed for the area around Cape Bolinao. This is called the Fine Resolution Model which is developed independently from the Coarse Model.

4.1. Theoretical Considerations

Fundamental to understanding sediment transport and light extinction processes that occur at the coastal sea is a thorough knowledge on coastal hydrodynamics. The coastal sea features some unique physical characteristics not present in many natural surface waters. Firstly, the long gravity wave known as tide, whose propagation from the open sea to the coast, presents some interesting characteristics not observed in many freshwater systems. The resulting water level variation and tidal current patterns dictate many observable phenomena at the coast which include sediment and nutrient transport processes. Secondly, the high salinity content of marine waters as compared to freshwater adds complexity to the density structure in both horizontal and vertical directions.   In particular, the salinity variations near rivers enhance stratification in the water column which can have significant influence not only on hydrodynamics but also on transport processes, mixing and entrainment in the water column.

The necessary coastal hydrodynamical principles needed in this study are presented in the following sections.

4.1.1.   Tide-Induced Coastal Circulation

The coastal circulation is dictated in part by the tide-generating forces producing a periodic ebbing and flooding near the coast. The tide is simply a long gravity wave with wavelength scaled in kilometers, and periods from several hours to days. The typical speed of propagation approximately equals $(gh)^{1/2}$, where g is the gravitational acceleration and h is the water depth. This is just the speed of waves propagating in a shallow sea. Comparing the wavelength L, which is several hundreds of kilometers, to the water depth h ($h < 1000$ m), a tide a can be categorized as a shallow water wave ($L/h > 25$).

The Lingayen Gulf has been characterized by the dominance of the four tidal constituents O_1, K_1, M_2, and S_2. A typical tidal observation in the gulf reveals a mixed tide in the basin with a dominant diurnal component. The gulf, being oriented approximately north-south in the west coast of Luzon with its opening to the north, possesses a distinct tidal circulation. Tide-induced currents in such a basin would reveal rotating north-south components of flow velocities. Currents would rush towards the south during flooding, and towards the north during ebbing. This simple flow pattern is however modified and complicated by the wind stress acting over the area. The wind stress and tide forces interact non-linearly to produce a complex flow pattern partly modified by the coastal geometry, the bathymetric configuration and presence of islands within the gulf.

The tide-induced motion of a coastal sea is generally described by the conservation laws of momentum and water mass (derivation is beyond the scope of this thesis). In a 2-dimensional cartesian coordinate system, the tidal movement in a coastal sea can be described by the basic equations:

$$\frac{\partial u}{\partial t} + u\frac{\partial u}{\partial x} + v\frac{\partial u}{\partial y} = fv - g\frac{\partial \zeta}{\partial x} + \frac{ku\sqrt{u^2 + v^2}}{h} + v_h\nabla_h^2 u \qquad (4.1)$$

$$\frac{\partial v}{\partial t} + u\frac{\partial v}{\partial x} + v\frac{\partial v}{\partial y} = -fu - g\frac{\partial \zeta}{\partial y} + \frac{kv\sqrt{(u^2 + v^2)}}{h} + v_h\nabla_h^2 v \qquad (4.2)$$

$$\frac{\partial \zeta}{\partial t} + \frac{\partial(uh)}{\partial x} + \frac{\partial(vh)}{\partial y} = 0 \qquad (4.3)$$

where u and v represent the depth-averaged current velocities defined as

$$u = \frac{1}{h}\int_{-h_0}^{\zeta} u\,dz \quad , \quad v = \frac{1}{h}\int_{-h_0}^{\zeta} v\,dz \qquad (4.4)$$

and ζ is the sea surface elevation. Earth's rotation is taken into consideration by the coriolis parameter f in Equations (4.1) and (4.2) defined as

$$f = 2\Omega\sin(\Phi) \qquad (4.5)$$

where Ω is the angular rotation rate of the earth ($\sim 7.292 \times 10^{-5}$ rad s^{-1}) and Φ is the latitude. The time-dependent water level h is defined as

$$h = h_o + \zeta \qquad (4.6)$$

where h_o is the still water depth.

Conservation of momentum is represented by Equations (4.1) and (4.2). Together, they describe the horizontal balance of forces per unit mass in an incompressible fluid.

Specifically, these equations describe the evolution of the flow velocities at a particular location. The mean flow is dictated by momentum advection, coriolis acceleration, surface elevation gradient, bottom frictional dissipation and horizontal momentum diffusion (in the order written).

The conservation of water mass is simply written in Equation (4.3). Basically called the depth-integrated equation of mass continuity, the evolution of the sea surface elevation due to the tide is described by this equation.

The quadratic formulation of the bottom friction term (second to the last terms in Equation 4.1 and 4.2) is a rather arbitrary option. Bottom friction can be parameterized as a linear function of the flow velocity. In cases where quadratic parameterization is used, the parameter k is often taken to be related to the Chezy coefficient C as in

$$k = \frac{g}{C^2} \qquad (4.7)$$

where g is the acceleration due to gravity.

The tide is usually described by a perturbation coming from the open sea boundary propagating inside a bay where tidal circulation is being modelled. The simultaneous solution of the system of partial differential equations (4.1 to 4.3) describes the evolution of the horizontal current velocities and surface elevation fields in a coastal sea due to the tide. Neglecting diffusion represented by the horizontal Laplacian terms (last terms in 4.1 and 4.2), Flather and Heaps (1975) successfully modelled the tidal dynamics of the Morecambe Bay using a finite difference numerical model based on the preceding non-linear equations.

4.1.2. Wind-Driven Coastal Circulation

The wind exerts a stress on the sea surface which is approximately proportional to the square of its speed. The general circulation pattern in a coastal sea is complicated by the time-dependent wind stress. This stress is also variable in space, when considering large coastal seas, contributing to a complex flow pattern governed by the interaction of the wind-driven and the tide-induced currents. The shallow depths in coastal waters makes it possible for the wind to dominate over the tide in some occasions. It is generally accepted that the wind stress effect is important in areas with depths of about 100 m or less. These are typically the areas within a continental shelf, from the coast to several kilometers to the open sea. In such areas, it is therefore almost impossible to decouple the effect of the wind and tide in the general circulation pattern.

Generally, the basic conservation equations (4.1 - 4.3) also apply to the wind-driven circulation in a coastal sea with the inclusion of an appropriate surface stress forcing from the wind. Koutitas (1988) gives the classical model for wind-driven circulation in a coastal sea as

$$\frac{\partial u}{\partial t} + u\frac{\partial u}{\partial x} + v\frac{\partial u}{\partial y} = fv - g\frac{\partial \zeta}{\partial x} + \frac{\tau_{sx} - \tau_{bx}}{\rho h} \qquad (4.8)$$

$$\frac{\partial v}{\partial t} + u\frac{\partial v}{\partial x} + v\frac{\partial v}{\partial y} = -fu - g\frac{\partial \zeta}{\partial y} + \frac{\tau_{sy} - \tau_{by}}{\rho h} \qquad (4.9)$$

$$\frac{\partial \zeta}{\partial t} + \frac{\partial (uh)}{\partial x} + \frac{\partial (vh)}{\partial y} = 0 \qquad (4.10)$$

where the τ_{sx} and τ_{sy} represent the surface stress due to the wind, τ_{bx} and τ_{by} the bottom stress, and ρ is the water density. The surface stress is generally parameterized in terms of the wind W taken at anemometer level as in

$$\tau_{sx} = \rho_a c_d W_x |W| \qquad , \qquad \tau_{sy} = \rho_a c_d W_y |W| \qquad (4.11)$$

with a constant or variable drag coefficient c_d. For example, Wu (1982) gives

$$c_d = (0.8 + 0.065\,W) \times 10^{-3} \qquad (4.12)$$

from breeze to hurricane with the wind speed in m/s. On the other hand, the bottom stress is generally represented by a quadratic relation as in

$$\frac{\tau_{bx}}{\rho} = ku\sqrt{(u^2 + v^2)} \qquad , \qquad \frac{\tau_{by}}{\rho} = kv\sqrt{(u^2 + v^2)} \qquad (4.13)$$

where k is the bottom friction coefficient which can be estimated in terms of the Chezy coefficient. Similar equations can be used in modelling lake circulation. For example, Van Duin (1992) applied the model WAQUA where horizontal transport by flow velocities in the shallow Lake Marken (The Netherlands) is modelled on the basis of the preceding non-linear equations with the inclusion of horizontal momentum diffusion.

The sea surface elevation ζ defined by the wind has a wavelength scaled in kilometers and a period of a few hours. The general wind-driven circulation develops within a period of several hours to days.

The wind-driven coastal circulation described by Equations (4.8 - 4.10) is responsible for the long-term and long-distance transport of suspended particulate. Normally, fine particles with very low settling velocities remain suspended in the water column. The wind-driven flow velocities are generally responsible for the horizontal transport and redistribution of these fine materials. The transport of larvae and other planktonic materials in coastal seas and lakes is governed by the drift currents due to the wind. For these reasons, the wind-driven component of water motion is of utmost ecological importance.

4.1.3. Surface Waves

Aside from the long-wave phenomenon defined by the wind as described in the previous section, surface waves with wavelengths scaled in meters and periods of several seconds can be observed in the coastal sea as perturbations superimposed on the tide and wind-induced surface elevation. The random behavior of the sea surface during moderate to strong wind events is defined by these surface waves.

Wind-waves as they are often called, are generally characterized by the significant wave height H_s, the significant wave period T_s, and the significant wavelength L_s. They are defined as the average height, period and wavelength of the one third highest waves. Several researches have dealt with the study of wind-waves. Groen and Dorresteyn (1976), Bouws (1986) and CERC (1977, 1984) provide empirical relationships for the determination of short-wave characteristics as induced by the wind.

The significant wave height H_s can be estimated from (CERC 1984, Bouws 1986);

$$H_s = \frac{0.283\,W^2}{g}\tanh\left[0.53\left[\frac{gh}{W^2}\right]^{0.75}\right]\tanh\left[\frac{0.0125\left[\frac{gF}{W^2}\right]^{0.42}}{\tanh\left[0.53\left[\frac{gh}{W^2}\right]^{0.75}\right]}\right] \qquad (4.14)$$

The significant wave period is similarly estimated from;

$$T_s = \frac{7.54\,W}{g}\tanh\left[0.833\left[\frac{gh}{W^2}\right]^{0.375}\right]\tanh\left[\frac{0.077\left[\frac{gF}{W^2}\right]^{0.25}}{\tanh\left[0.833\left[\frac{gh}{W^2}\right]^{0.375}\right]}\right] \qquad (4.15)$$

where F is the fetch or effective distance the wind W is blowing over a certain period. The wind is assumed by these formulations to be taken at anemometer level which is approximately at a height of 10 m above the water surface.

The wavelength can be estimated from the explicit relationship given by Fenton and McKee (1990);

$$L_s = \frac{gT_s^2}{2\pi}\left[\tanh\left(\frac{2\pi}{T_s}\sqrt{h/g}\right)^{3/2}\right]^{2/3} \qquad (4.16)$$

From the above formulations, it can be seen that the wave characteristics are highly dependent on the water depth, fetch, and wind speed. As the effective water depth is highly variable in the coastal marine environment due to the tide, an appropriate modelling of the tide is an important factor in short-wave hindcasting especially in shallow areas near the coast where the tidal fluctuation is a substantial portion of the total water depth. It should be noted that the surface waves estimated from these empirical relationships give average, instantaneous wave characteristics. From measurements taken in both lakes and coastal seas, these equations are generally reasonable approximations of the significant wave height, period and length respectively. They are particularly useful in estimating wave-induced bottom orbital velocities and near bed shear stress needed when considering rates of sediment resuspension, a topic for the next chapter.

4.2.   Modelling Coastal Circulation

Coastal circulation is generally a long-wave phenomenon primarily dictated by the simultaneous action of the wind and the tide. Modelling the circulation of a coastal area using the conservation laws of mass and momentum has been the general practice in coastal engineering and oceanography, and similar direction is adopted in this study.

The basis of the coastal hydrodynamical model developed in this study is the modified wind-driven circulation model of Koutitas (1988). It is extended in the present study to include horizontal momentum diffusion and some modifications in the general application. The governing equations read:

$$
\begin{aligned}
\frac{\partial u}{\partial t} &+ u\frac{\partial u}{\partial x} + v\frac{\partial u}{\partial y} + \left(0.2u + \frac{a_x}{40}\right)\frac{\partial u}{\partial x} + \left(0.2v + \frac{a_y}{40}\right)\frac{\partial u}{\partial y} = fv \\
&- g\frac{\partial \zeta}{\partial y} + \frac{\tau_{sx}}{\rho h} - \left(0.18\frac{u}{h}\sqrt{\tau_s/\rho} - 0.5\frac{\tau_{sx}}{\rho h}\right) + A_h\left(\frac{\partial^2 u}{\partial x^2} + \frac{\partial^2 u}{\partial y^2}\right)
\end{aligned}
\tag{4.17}
$$

$$
\begin{aligned}
\frac{\partial v}{\partial t} &+ u\frac{\partial v}{\partial x} + v\frac{\partial v}{\partial y} + \left(0.2u + \frac{a_x}{40}\right)\frac{\partial v}{\partial x} + \left(0.2v + \frac{a_y}{40}\right)\frac{\partial v}{\partial y} = -fu \\
&- g\frac{\partial \zeta}{\partial y} + \frac{\tau_{sy}}{\rho h} - \left(0.18\frac{v}{h}\sqrt{\tau_s/\rho} - 0.5\frac{\tau_{sy}}{\rho h}\right) + A_h\left(\frac{\partial^2 v}{\partial x^2} + \frac{\partial^2 v}{\partial y^2}\right)
\end{aligned}
\tag{4.18}
$$

$$
\frac{\partial \zeta}{\partial t} + \frac{\partial (uh)}{\partial x} + \frac{\partial (vh)}{\partial y} = 0
\tag{4.19}
$$

where τ_s represents the stress acting over the water surface, f is the Coriolis parameter and A_h is the horizontal eddy viscosity coefficient which can be assumed constant. The variable a is related to the wind stress as in:

$$
a = \frac{\tau_s h}{\rho \nu}
\tag{4.20}
$$

The eddy viscosity ν is assumed constant with mean value equivalent to

$$\bar{\nu} = \lambda h \sqrt{\tau_s/\rho} \tag{4.21}$$

Turbulence models and laboratory measurements indicate that $O[\lambda] = 0.1$ (Koutitas 1988). For $\lambda = 0.066$, Koutitas gives

$$a = \frac{\tau_s h}{\rho \nu} = 16.6 \sqrt{\tau_s/\rho} \tag{4.22}$$

where ρ is the water density. The surface stress is assumed as a quadratic function of the wind as given in Equation (4.11).

Equations (4.17) and (4.18) define the current accelerations in the horizontal x and y-directions respectively. They are basically a representation of the interactions of the physical processes inducing the motion of a water body subject to known forcing functions from wind stress and tide. The first terms on the left of both equations represent the local change of the flow velocities. The following terms on the left represent changes in the fluid acceleration due to advection of momentum. The additional advective terms involving the stress variable a are corrections imposed on advection to include non-uniformity in the current profile (see Koutitas 1988). On the right hand side of both equations (in the order written), effects due to earth's rotation (coriolis acceleration), horizontal elevation gradient, surface stress and bottom frictional effects, and horizontal momentum diffusion provide the necessary physical processes for modelling coastal circulation. Basically, they represent conservation of momentum in the coastal sea. Current velocities are predicted using these equations. On the other hand, Equation (4.19) which is simply the continuity equation, represents conservation of water mass. It predicts the evolution of water levels or surface elevations from known current velocities due to the wind and the tide.

These equations, while written in a 2-dimensional vertically-integrated form, can be used to assess the three dimensional structure of the water current. It is developed by assuming that the current profile in the vertical is a quadratic function of depth i.e. $u(z) = az^2 + bz + c$. With appropriate boundary conditions, Koutitas (1988) derived the solution for the current profile in the vertical to be:

$$u(z) = \left(\frac{3}{4}a - \frac{3}{2}u\right)\left[\left(\frac{z}{h}\right)^2 - 1\right] + a\left(\frac{z}{h} + 1\right) \qquad (4.23)$$

The hydrodynamic model is thus a quasi-three dimensional model in the sense that currents at any depth can be estimated from model calculations using Equation (4.23).

The effect of the tide in the coastal circulation can be included by the propagation of a long-gravity wave at the open boundary of the computational domain. The surface elevation field, varying in time, is normally prescribed at the open sea boundary with an appropriate tidal forcing function representative of the observed tidal variation within the area of interest.

In the application of the general Equations (4.17 - 4.19), it should be noted that the bottom friction has to be modified in case the wind is very low or approaches zero. Frictional dissipation would disappear in such cases as dictated by the modified bottom friction

$$\frac{\tau_{bx}}{\rho} = 0.18u\sqrt{\tau_s/\rho} - 0.5\frac{\tau_{sx}}{\rho} \quad , \quad \frac{\tau_{by}}{\rho} = 0.18v\sqrt{\tau_s/\rho} - 0.5\frac{\tau_{sy}}{\rho} \qquad (4.24)$$

With the tide dominating the flow in such instance, this is unrealistic and would tend to give unstable solutions due to high velocity fluctuations. In the coastal sea, bottom frictional dissipation never disappears and in cases when the wind approaches zero, the standard quadratic friction law given by Equation (4.13) should be used. This is an additional modification to the original modified circulation model of Koutitas (1988) devised mainly to include the non-linear interaction between the wind and the tide.

4.2.1. Finite Differentiation and Numerical Integration

It is necessary to solve the three partial differential equations (4.17 to 4.19) simultaneously in order to assess the water movement due to the primary forcing functions. Modelling the circulation in a coastal sea using these governing equations (or simplified versions with reduced number of terms) is not an easy task. The numerical solution of the system of equations has been the subject of many researches in hydraulics and oceanography for the past decades. Many approaches based on finite difference and finite element methods have been proposed. The solution approach in this study follows that of Koutitas (1988). It is basically an explicit finite difference approach defined in a staggered grid shown in Figure

(4.1) with the indices i and j representing specific locations of a variable in the x and y-directions respectively. The basic grid element shown represents a fundamental area of the coastal sea.

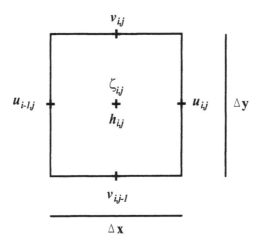

Figure 4.1. Location of variables in the staggered grid used in the present study.

In this grid system, the x-component of velocity u is situated at the middle of a y-directed side of the basic grid element shown in the figure. Also, the y-component of velocity v is situated at the middle of an x-directed side. The current velocities u and v are offset half a grid distance from the surface elevation ζ which is situated at the center of a grid cell. The water depth h is similarly defined at the center of the grid cell. In modelling applications, this space-staggered approach has a big advantage over other techniques in that boundaries are treated quite easily. For example, in solid coastal boundaries, vanishing normal components of flow velocities are easily specified. At the open boundaries, lesser number of variables need to be determined at each time during the numerical integration. Furthermore, the associated spatial gradients of the variables in each of the governing equations can also be treated with simple space-differences.

Assuming a finite grid distance of $\Delta x = \Delta y = \Delta s$ and a finite time interval Δt, the explicit finite difference representations of the general equations (4.17 - 4.19) in a staggered grid shown in Figure (4.1) are given by

$$u_{i,nj}^{n+1} = u_{i,nj}^{n} - \Delta t[(1.2u_{i,j}^{n} + 0.4sgn(\tau_{sx}/\rho)\sqrt{|\tau_{sx}/\rho|})(\frac{u_{i+1,j}^{n} - u_{i-1,j}^{n}}{2\Delta s})$$

$$+ (1.2\bar{v}^{n} + 0.4sgn(\tau_{sy}/\rho)\sqrt{|\tau_{sy}/\rho|})(\frac{u_{i,j+1}^{n} - u_{i,j-1}^{n}}{2\Delta s})$$

$$+ g(\frac{\zeta_{i+1,j}^{n+1} - \zeta_{i,j}^{n+1}}{\Delta s}) - f\bar{v}^{n} - \frac{\tau_{sx}}{\rho \bar{h}_{x}} + (0.18\frac{u_{i,j}^{n}}{\bar{h}_{x}}\sqrt{(\tau_{s}/\rho)} - 0.5\frac{\tau_{sx}}{\rho \bar{h}_{x}})$$

$$- A_{h}(\frac{u_{i+1,j}^{n} + u_{i-1,j}^{n} - 4u_{i,j}^{n} + u_{i,j+1}^{n} + u_{i,j-1}^{n}}{\Delta s^{2}})]$$

(4.25)

$$v_{i,nj}^{n+1} = v_{i,nj}^{n} - \Delta t[(1.2v_{i,j}^{n} + 0.4sgn(\tau_{sy}/\rho)\sqrt{|\tau_{sy}/\rho|})(\frac{v_{i,j+1}^{n} - v_{i,j-1}^{n}}{2\Delta s})$$

$$+ (1.2\bar{u}^{n} + 0.4sgn(\tau_{sx}/\rho)\sqrt{|\tau_{sx}/\rho|})(\frac{v_{i+1,j}^{n} - v_{i-1,j}^{n}}{2\Delta s})$$

$$+ g(\frac{\zeta_{i,j+1}^{n+1} - \zeta_{i,j}^{n+1}}{\Delta s}) + f\bar{u}^{n} - \frac{\tau_{sy}}{\rho \bar{h}_{y}} + (0.18\frac{v_{i,j}^{n}}{\bar{h}_{y}}\sqrt{(\tau_{s}/\rho)} - 0.5\frac{\tau_{sy}}{\rho \bar{h}_{y}})$$

$$- A_{h}(\frac{v_{i+1,j}^{n} + v_{i-1,j}^{n} - 4v_{i,j}^{n} + v_{i,j+1}^{n} + v_{i,j-1}^{n}}{\Delta s^{2}})]$$

(4.26)

$$\zeta_{i,j}^{n+1} = \zeta_{i,j}^{n} - \frac{\Delta t}{2\Delta s}[u_{i,j}^{n}(h_{i+1,j} + h_{i,j}) - u_{i-1,j}^{n}(h_{i,j} + h_{i-1,j}) + v_{i,j}^{n}(h_{i,j+1} + h_{i,j})$$

$$- v_{i,j-1}^{n}(h_{i,j} + h_{i,j-1})]$$

(4.27)

where the average values of the current velocities u and v are given by;

$$\bar{u}^{n} = \frac{1}{4}(u_{i,j}^{n} + u_{i,j+1}^{n} + u_{i-1,j+1}^{n} + u_{i-1,j}^{n})$$

(4.28)

and

$$\bar{v}^{n} = \frac{1}{4}(v_{i,j}^{n} + v_{i,j-1}^{n} + v_{i+1,j-1}^{n} + v_{i+1,j}^{n})$$

(4.29)

The discrete values of the variables are denoted by the indices i and j (x and y-axes respectively), with the time level denoted by the superscript n. As a result of the space staggered approach, the values of u at a v-point, and the values of v at a u-point are taken as simple averages of four neighboring points (Equations 4.28 - 4.29). Also, the mean water depths used at the locations of the x and y-velocity components are respectively given by

$$\overline{h}_x = \frac{1}{2}(h_{i+1,j}^n + h_{i,j}^n) \quad , \quad \overline{h}_y = \frac{1}{2}(h_{i,j+1}^n + h_{i,j}^n) \tag{4.30}$$

where the effective water depth $h = h_o + \zeta$, at a particular time level n.

The surface stress terms τ_{sx}, τ_{sy} and τ_s are time-dependent but nevertheless do not depend on locations since it is assumed that the wind does not vary considerably within the computational domain. This is a reasonable assumption considering the small spatial scale of the area of application (< 100 km). Spatial gradients in the wind stress is appreciable on a space scale of greater than 100 km (Gill 1982).

The mathematical modelling problem is to provide the solution of the three unknown variables u, v and ζ as functions of space and time. The numerical solution proceeds by solving Equation (4.27) first providing new ζ values at time $n+1$. Using this updated surface elevation field, calculation of the new velocity field from Equations (4.25) and (4.26) follows providing predicted values of u and v at time level $n+1$. Calculations always proceed in the order of increasing i and j respectively.

Initial condition assumes that the coastal sea is at rest which means that $u(x,y,t)$, $v(x,y,t)$ and $\zeta(x,y,t) = 0$ at $t = 0$. After one day of simulation, the modelled area generally becomes insensitive to the effects of initial conditions due to frictional dissipation. Furthermore, the modified model with corrections to horizontal dispersion and bottom friction, tends to converge (reach steady state) faster than the classical one given in Equations (4.8 - 4.10) (Koutitas 1988). This makes it more suitable to use the new non-linear model to assess the time evolution of the flow velocities and surface elevation fields in the coastal sea subject to the wind and tidal forcing.

It should be noted that in the numerical solution of the system of equations (4.25 to 4.27), the advection and diffusion terms have to be solved a grid distance Δs from the open boundary. Unlike in the non-linear model of Flather and Heaps (1975) where advection is decoupled $3\Delta s$ from the boundary to remove the effect of boundary conditions from the non-linear terms, the present model when used in conjunction with a suitable form of the Orlanski Radiation Condition permits advection and diffusion only a grid distance from the open boundary (see Section 4.3.2). This is physically more realistic in the sense that momentum is allowed to be advected and diffused at the boundary, a restriction imposed in other modelling studies. It is likewise worth mentioning the importance of the time-variation of the total water depth h in costal modelling. In shallow areas for example, the bottom friction terms contain a singularity when the depth approaches zero at low tide. Therefore, in the general application of the model, h is set to 1 m whenever the effective depth $(h_o + \zeta)$ becomes lower than this minimum value.

4.3. Coarse Resolution Model - The Lingayen Gulf Case

The hydrodynamic and transport phenomena at the Lingayèn Gulf determine in part the dynamics of the main area of interest which is the Cape Bolinao. This is one important reason of modelling the whole Lingayen Gulf. The big rivers discharging south of the gulf can have some siltation impact at the Cape Bolinao reef system. For example, slow-settling, fine particulate materials discharged by these rivers which may be nutrient-rich and carrying some contaminants can reach Cape Bolinao by means of the advective currents driven by the wind and the tide. To answer such questions, it is the necessary to investigate through numerical modelling the overall dynamics of the Lingayen Gulf.

The Coarse Resolution Model applied at the Lingayen Gulf is based on the modified non-linear equations (4.17 - 4.19). The finite difference equations approximating the solutions of these partial differential equations are given by equation (4.25 to 4.27). Section (4.2.1) describes the numerical integration of these equations. The following sections give additional details on the modelling procedure.

4.3.1. Grid Layout and Bathymetric Smoothing

The computational area for the Coarse Model covers the whole of Lingayen Gulf. For reasons of computational accuracy, extension of the domain from the mouth of the gulf to several kilometers of the open sea is given allowance. Figure (4.2) shows the layout of the computational grid defining the coastal boundaries and several islands covered by the grid network. The grid distance Δs is 1 km in the Coarse Model.

At the center of each grid cell, the water depth is read from available bathymetric maps published by the Bureau of Coast and Geodetic Survey (BCGS Philippines, 1984). The datum level used in these maps is the mean lower low water (MLLW). The depth distribution in the Lingayen Gulf is rather complex. For modelling purposes, smoothing is applied using

$$h_{i,j} = sh_{i,j} + (1 - s)\frac{[h_{i+1,j} + h_{i,j+1} + h_{i-1,j} + h_{i,j-1}]}{4} \qquad (4.31)$$

where s is a smoothing coefficient between 0 and 1. A value of 0.25 is used in the model.

North Open Boundary

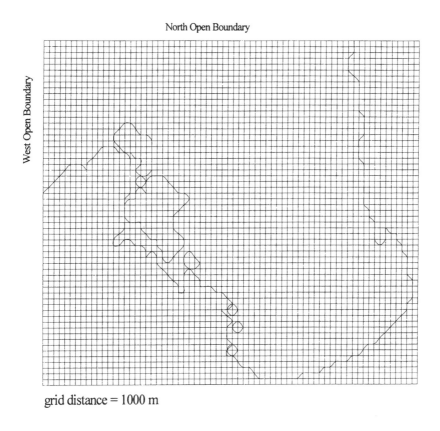

grid distance = 1000 m

Figure 4.2. Grid layout for the Coarse Resolution Model.

Figure (4.3) shows a three-dimensional picture of the depth distribution at the Lingayen Gulf smoothed according to Equation (4.31). The decrease in depth distribution is quite abrupt towards the west and northwest of Cape Bolinao. Depths in that region reach more than 500 m. In modelling long gravity waves, great depths would result in very short time steps to maintain stable computations since the time step, Δt, in model integration is determined by:

$$\Delta t \; < \; \frac{\Delta s}{\sqrt{2gh_{max}}} \tag{4.32}$$

Better known as the Courant-Friedrichs-Lewy (CFL) criterion for computational stability, this implies that the time interval for modelling long wave circulation should be less than the time

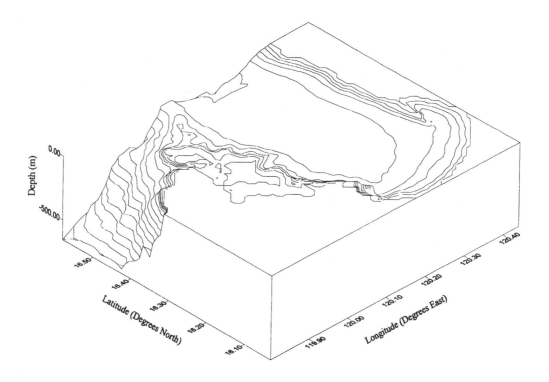

Figure 4.3.   Depth distribution at the Lingayen Gulf.

needed for the long gravity wave to cover the grid distance Δs. Numerical instability can result if the time interval exceeds this value. With a grid distance Δs of 1000 m and a maximum depth $h_{max} > 500$ m, the optimum time interval should be less than 10 s. Since the present study is concerned on localized processes at the coastal zone, and taking advantage of the fact that the wind stress is important at depths of about 100 m, the time interval for numerical simulation can be put to 10 s when the maximum depth is set to 100 m. Using such Δt generally gives stable solutions for the present coastal wind and tide-driven) circulation model.

4.3.2. Open Boundary Conditions: The Pseudo-Implicit Orlanski Radiation Condition

For more realistic simulations in a limited-area domain, the boundaries at the open sea of the computational domain have to be modelled correctly. Unlike in the coastal boundaries where flows can be easily specified by zero-water transport, flows at the open sea boundaries are not easily predicted. Water transport in and out of the open boundaries during tidal flooding and ebbing has to be accurately specified in order to get an accurate prediction of the flows inside the computational domain.

There are two open boundaries for the coarse resolution model, in the north and west of the Lingayen Gulf (see Figure 4.2). These represent considerable difficulty in modelling tide and wind-driven circulation in the gulf. An incorrect representation of the flow velocity or water level at these open boundaries will result to inaccurate predictions near the boundaries and inside the computational domain. Errors at those locations can propagate inside and corrupt the predictions thereby giving unreliable computational results.

Several open boundary conditions are available in literature. For example, a Neumann type boundary condition wherein the spatial gradient of the variable normal to the boundary is assumed to vanish implying quasi-uniformity of the current velocity across the boundary. This is a simple yet efficient type of open boundary condition which can be used in coastal circulation modelling (Koutitas 1988).

Open boundary conditions based on the conservation of water mass given by the equation of mass continuity (Equation 4.27) is sometimes used. In modelling the tide-driven circulation of Morecambe Bay, Flather and Heaps (1975) applied this type of open boundary condition and successfully predicted the dynamics of the basin as forced by the M_2 tide.

In a series of numerical experiments, Chapman (1985) proposes several types of open boundary condition which can be used in coastal ocean modelling. Ranging from a clamped boundary condition (no change in time of a variable at the boundary), Neumann type, and several radiation conditions (wave propagation), the reflection coefficient of each of the studied open boundary condition is estimated. The result of the numerical experiments showed that a special case of the Orlanski radiation condition appears to be a 'perfect absorber' of perturbations generated inside the computational domain, i.e. zero-reflection. The use of such open boundary condition is thus favorable.

The Orlanski Radiation Condition is basically a wave-propagation technique whereby the value of a variable at the open boundary can be determined from the inner neighboring values if the propagation speed of that variable is known. The approach is generally based on the Sommerfeld radiation condition represented by the advection equation (Chapman 1985);

$$\frac{\partial \phi}{\partial t} \pm c \frac{\partial \phi}{\partial x} = 0 \tag{4.33}$$

where ϕ represents either the velocity or sea surface elevation, and c is the advective velocity or phase speed. Note that the upper and lower sign corresponds to the right and left open boundaries respectively. This type of open boundary condition gives a time-dependent approximation of a variable if certain conditions for the propagation speed of that variable are satisfied near the boundary. These necessary conditions defining the propagation speed c of a variable ϕ are given by;

$$c = \begin{cases} \dfrac{\Delta x}{\Delta t} & if \quad \mp \dfrac{\partial \phi / \partial t}{\partial \phi / \partial x} \geq \dfrac{\Delta x}{\Delta t} \\[2ex] \mp \dfrac{\partial \phi / \partial t}{\partial \phi / \partial x} & if \quad 0 < \mp \dfrac{\partial \phi / \partial t}{\partial \phi / \partial x} < \dfrac{\Delta x}{\Delta t} \\[2ex] 0 & if \quad \mp \dfrac{\partial \phi / \partial t}{\partial \phi / \partial x} \leq 0 \end{cases} \tag{4.34}$$

In its numerical form, Chapman (1985) recommends an implicit version of Equations (4.33-4.34). This is basically a pseudo-implicit form as boundary values are estimated partly from already known, predicted quantities. Values of a variable at and near the open boundary at time levels $n+1$, n and $n-1$ are used in the time and space derivatives of ϕ. The original Orlanski Radiation condition is an explicit formulation requiring only quantities at time levels n, $n-1$, and $n-2$ to obtain the boundary variable at time level $n+1$. These can be seen more clearly in the finite difference analogue of Equation (4.33) which can be written as

$$\frac{\phi_B^{n+1} - \phi_B^{n-1}}{2 \Delta t} \pm c \left[\frac{(\phi_B^{n+1} + \phi_B^{n-1})/2 - \phi_{B \mp 1}^n}{\Delta x} \right] = 0 \tag{4.35}$$

where the subscript B denotes the boundary. Note that in the advective term, the boundary value of the variable is taken as the mean value at two time-levels (i.e. at $n+1$ and $n-1$) while the value at a gridpoint next to the boundary is taken at time level n. Assuming a dimensionless quantity, μ, (sometimes called the Courant number) given by

$$\mu = c\frac{\Delta t}{\Delta x} \tag{4.36}$$

then Equation (4.34) is transformed to;

$$\mu = \begin{array}{lll} 1 & if & C_L \geq 1 \\ C_L & if & 0 < C_L < 1 \\ 0 & if & C_L \leq 0 \end{array} \tag{4.37}$$

where C_L is given by

$$C_L = \frac{\phi_{B\mp1}^{n-1} - \phi_{B\mp1}^{n+1}}{\phi_{B\mp1}^{n+1} + \phi_{B\mp1}^{n-1} - 2\phi_{B\mp2}^{n}} \tag{4.38}$$

Note that the time and space derivatives of the next inner grid points are used to obtain C_L. This implies that it is only necessary to get information on the propagation speed c (and consequently of μ) of a variable ϕ at the next inner grid points in order to know the value of that variable at the open boundary. Rearranging Equation (4.35) and substituting μ for $c\Delta t/\Delta x$, we finally obtain;

$$\phi_B^{n+1} = [\phi_B^{n-1}(1 - \mu) + 2\mu\phi_{B\mp1}^{n}]/(1 + \mu) \tag{4.39}$$

where ϕ_B^{n+1} is the predicted value of a variable ϕ (either velocity or sea surface elevation) at the boundary. The position of the variable at the open boundary defines the sign to be used in the implementation of Equations (4.37 - 4.39) in the model i.e. the upper sign corresponds to the right open boundary and the lower sign corresponds to the left open boundary. While the Coarse Resolution model presently discussed needs only the right open boundary formulations, the Fine Resolution model (section 4.4) needs both boundary formulations.

Application of the above boundary condition to the normal component of current velocity at the western open boundary is straightforward. Equation (4.39) gives for the u-component of velocity;

$$u_{0,j}^{n+1} = [u_{0,j}^{n-1}(1 - \mu) + 2\mu u_{1,j}^{n}]/(1 + \mu) \qquad (4.40)$$

where $u_{0,j}$ is the estimated open boundary value of the current at the west boundary. The value of μ has to be estimated according to Equations (4.37). The necessary conditions satisfying (4.37) can be estimated from:

$$C_L = \frac{u_{1,j}^{n-1} - u_{1,j}^{n+1}}{u_{1,j}^{n+1} + u_{1,j}^{n-1} - 2u_{2,j}^{n}} \qquad (4.41)$$

Specification for the current component at the north open boundary is also needed for calculation of the non-linear terms. The procedure is also based on the same radiation condition and this gives;

$$v_{i,nj}^{n+1} = [v_{i,nj}^{n-1}(1 - \mu) + 2\mu v_{i,nj-1}^{n}]/(1 + \mu) \qquad (4.42)$$

where the subscript nj (maximum j) refers to the north open boundary value. The estimation of the phase speed requires that the next two inner grid points at the north open boundary be used. Thus, we obtain for C_L

$$C_L = \frac{v_{i,nj-1}^{n-1} - v_{i,nj-1}^{n+1}}{v_{i,nj-1}^{n+1} + v_{i,nj-1}^{n-1} - 2v_{i,nj-2}^{n}} \qquad (4.43)$$

The surface elevation field at the open boundary denoted by ζ is specified as a tide propagating inside the computational domain. The following section describes the derivation and application of a tidal boundary condition at the northern open boundary.

4.3.3.   External Tidal Forcing

The north open boundary condition for surface elevation ζ is specified using the Fourier analyzed tide at Cape Bolinao. Continuous time series of hourly tide level observed at

Bolinao are used for the analysis. The dominant tidal constituents O_1, K_1, M_2, S_2 are incorporated using their known frequencies. (Table II.3 gives the periods of the dominant tidal constituents in the area.) The result of the analysis provides a suitable tidal forcing function to be used in this study. The tidal variation described by such forcing function is given by the equation:

$$\zeta(t) = a_o + a_1 \cos(\omega_1 t - p_1) + a_2 \cos(\omega_2 t - p_2) + a_3 \cos(\omega_3 t - p_3)$$
$$+ a_4 \cos(\omega_4 t - p_4) \quad (4.44)$$

where $\zeta(t)$ represents the sea surface elevation due to the tide at the open sea boundary as a function of time t, a's the coefficients or tidal amplitudes, ω's their known frequencies, and p's the phases of each of the four constituents.

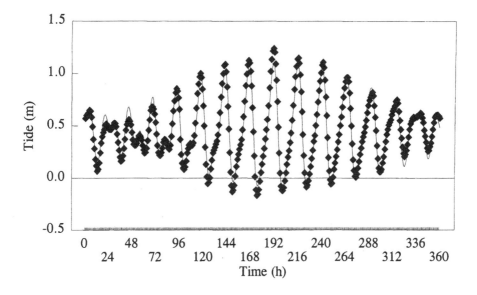

Figure 4.4. Tidal forcing derived from the Fourier-analyzed data at Cape Bolinao.

The truncated Fourier series given by this equation is then used to force the model at the northern open boundary simulating periodic flooding and ebbing at the Lingayen Gulf. Figure (4.4) shows a typical tidal variation based on the Fourier equation (4.44) whose tidal constants (amplitudes and phases) are derived from the observed tide (also shown) at the study area.

While the tidal variation shows a typical diurnal characteristic showing one high and one low tide level in a day, an occasional semi-diurnal character of the tide is observed. The interaction of the four tidal constituents produce this typical tidal curve in the study area with the diurnal components dominating over the semi-diurnal components.

4.4.   Fine Resolution Model - The Cape Bolinao Case

Cape Bolinao has a smaller space scale as compared to the whole Lingayen Gulf requiring higher resolution to understand fully well the dynamics occurring in that open coastal system. The Fine Resolution Model described in this study provides the modelling approach which can be used in such a small scale system. The approach is simply based on the same finite difference model used in the Coarse Resolution Model. Treatment of open boundaries is similar to the coarse model. However, this now requires more velocity components at the open boundaries. Following the same Orlanski Radiation Condition as discussed in Section (4.3.2), the application is nevertheless straightforward. Also, the use of the same tidal forcing as in the coarse model is applied to obtain similar forcing mechanisms for both modelled basins.

4.4.1. Grid Layout

The primary area of interest covers the reef flat around Santiago Island at Cape Bolinao. This is the region where most of the observations were made. The Fine Resolution Model covers all of these areas extending several kilometers from the reef to the open sea. For reasons of numerical accuracy, the effect of open boundaries has to be decoupled from the areas of interest.

The grid network can be seen in Figure (4.5) with the smoothed bathymetry in Figure (4.6). With the present resolution of 500-m grid distance, there are still small islands and channels between islands which, unfortunately, can not be represented in the model. Also, there are sub-grid scale intertidal areas which are exposed during low tide, and are submerged again during high tide. Nevertheless, most of the important features of the study area which include the Santiago Island and the channel systems surrounding Cape Bolinao are resolved quite well.

Similar to the coarse resolution model for Lingayen Gulf, the flow velocity and surface elevation field are defined using a space staggered technique as shown in Figure (4.1). This has the same obvious reason that boundaries are easy to handle in this grid type, i.e. the coastal boundaries do not permit normal flows, and open sea boundaries require less number of variables to be determined at any time level. These obvious advantages of the space

North Open Boundary

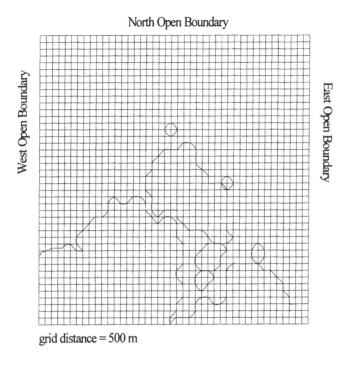

grid distance = 500 m

Figure 4.5.   Grid layout for the Fine Resolution Model.

staggered approach is the reason why many numerical coastal models, and also lake models rely on its use.   Van Duin (1992) described a similar approach for modelling the wind-driven circulation of Lake Marken in the Netherlands.   It is further claimed that the space staggering approach reduces computation time as the number of computational grid points, including the number of boundary values to be determined are reduced.

4.4.2. Treatment of Open Boundaries

Open boundaries are undesirable in any modelling work but nevertheless have to be approximated for the variables needed.   The boundary is an imaginary line dividing the modelled area from the rest of the sea.   The difficulty in modelling open coastal systems appear in the open boundaries where the numerical grid ends.   The propagation of quantities inside the computational domain across that boundary presents an interesting problem in coastal modelling.   If in cases where inappropriate boundary conditions are used in the model for example, reflections at the boundaries will result to high fluctuations in the regions near the boundary. This, if left unchecked, will corrupt the solution inside the computational domain, giving erroneous solutions.

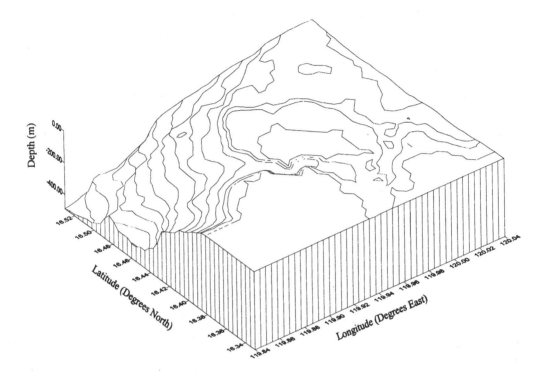

Figure 4.6.   Bathymetric chart of Cape Bolinao as used in the Fine Resolution Model.

A desirable boundary condition is one that is non-reflecting or weakly-reflecting (Vreugdenhil 1985, Chapman 1985). Ideally, there should be no reflection at all so that inside solutions can not be corrupted. The use of the Orlanski radiation condition partly achieves such situation at the open boundaries (Chapman 1985). The application of this open boundary condition in the Fine Resolution Model where three open boundaries exist (Figure 4.5) is discussed in the following section.

4.4.2.1. Application of the Orlanski Radiation Condition

With the present grid set-up, there is a lesser number of variables to be determined at each of the three open boundaries of the fine model. For the west and east open boundaries, the model requires the *u*-component of velocity. The north open boundary requires the *v*-

component of velocity and the sea surface elevation. The surface elevation field at both east and west boundaries can be determined explicitly from the continuity equation. However, the north open boundary requires that the tide be specified and allowed to propagate towards the coast.

The west open boundary requires the u-component of velocity. The computational procedure as outlined in the Lingayen Gulf model (Equations 4.40 - 4.41) is similarly applied in this Fine Resolution Model. The east open boundary now requires the u-component of velocity to be estimated. In accordance with Equation (4.39), the boundary value of the normal component of velocity u takes the form:

$$u_{ni,j}^{n+1} = [u_{ni,j}^{n-1}(1 - \mu) + 2\mu u_{ni-1,j}^{n}]/(1 + \mu) \qquad (4.45)$$

with the μ values also taken according to Equations (4.37). C_L is now estimated using the next two inner grid points to the left of the boundary as in:

$$C_L = \frac{u_{ni-1,j}^{n-1} - u_{ni-1,j}^{n+1}}{u_{ni-1,j}^{n+1} + u_{ni-1,j}^{n-1} - 2u_{ni-2,j}^{n}} \qquad (4.46)$$

For the north open boundary, the normal component of velocity v also needs to be estimated because of the non-linear terms. Application of the present boundary formulation has been outlined in the Coarse Resolution Model (Equations 4.42 - 4.43). The specification of the surface elevation field (tidal forcing) at the northern open boundary is the only boundary condition left. A brief description of this is provided in the following sub-section.

4.4.2.2. Tidal Forcing at Northern Open Boundary

Similar to the Coarse Resolution Model used for the Lingayen Gulf, the tidal forcing at the north open boundary of the Fine Resolution Model is based on the Fourier analyzed tidal data at Cape Bolinao. Similar condition is assumed representative of the tide at the open boundary of the Finer Resolution Model. When considering the wavelength of the tide which is several hundreds or thousands of kilometers, that small distance between the open boundary of the Coarse Model and the Fine Model is insensitive to the same tidal prescription. Only the deformation that can be caused by bathymetric change may be of importance but nevertheless not significant enough due to the very small distance covered and the slowly varying depth distribution within the area north of the gulf.

The Fourier representation given in Equation (4.44) is therefore applied using the estimated amplitudes and phases of the four dominant tidal constituents namely O_1, K_1, M_2 and S_2, as observed at Cape Bolinao. A typical tidal curve used to force the computational domain is shown in Figure (4.4).

Chapter 5

Sediment Transport in the Coastal Sea

Sediment transport plays an important role in water quality. Firstly, the crucial role that suspended sediments impart to the attenuation of the available photosynthetically useful radiant energy has long been documented. Secondly, contaminants and nutrients are generally transported along with the sediments upon which they are adsorbed. The transportation by currents and waves of especially the fine-grained sediments has lend itself then as a central theme for investigation by aquatic scientists. Sediment transport also plays an important role in coastal engineering and water quality management. The design, construction and operation of sound coastal structures especially for navigational purposes rely heavily on a knowledge of the sediment transport processes. These obvious reasons have become the background of intensive research on sediment transport in the coastal sea by both marine scientists and engineers in recent years.

Sediment transport in the coastal sea is a complex physical phenomenon involving the interactions of several processes. The knowledge obtained from river sediment transport studies has become the starting point for research but is not, in a strict sense, adequate to characterize transport processes in the coastal sea. While transport processes in the river are mainly dictated by channel flow velocities, the problem is complicated in the coast by the presence of both waves and currents. Their non-linear interaction makes the coastal transport process more complex. Additional complication arises from the fact that sediments transported into the sea become unstable due to the high salinity content of the marine environment. In particular, the so-called double layer thickness surrounding the fine (cohesive) particles is compressed due to an increased concentration of ions (Van Leussen 1994). This gives rise to the complex and continuous process of flocculation (aggregation), a phenomenon not very well understood until now.

The purpose of this chapter is to provide a theoretical and modelling background on coastal sediment transport. Specifically, this chapter focuses on suspended sediment transport which has significant bearing on light extinction (see Chapter 6) and nutrient transport processes. However, a brief discussion on the bed load transport of sediments is provided since this mode of sediment transport is closely related to the suspended load transport. Effects on the sediment transport by both currents and waves are treated in Sections (5.1 - 5.2). A practical modelling procedure (applied to the study area) for suspended sediment transport is discussed. The present study proposes the use of an open boundary condition for suspended sediment transport modelling based on the concept of wave propagation. The modelling approach, together with the proposed open boundary condition which is essential in dealing with open marine environmental systems, is discussed in Section (5.3).

5.1. Modes of Sediment Transport

There are two different modes of sediment transport at the coastal sea i.e. suspended and bed load transport. Description of the two processes, especially for a mixture of sediment particles with respect to size, is often difficult because they are closely related. Some practical definitions for the two processes are presented in the following sub-sections.

5.1.1. Bed-Load Transport

This mode of sediment transport refers to the motion of sediment particles which are either sliding, rolling or saltating with regular intermittent jumps close to the bed. By definition, the bed load transport zone where such sediment motion occurs is confined to a narrow layer near the bed assumed to be a multiple of the sediment grain diameter. Quantitative relationships for the bed load transport rate are given in Koutitas (1988) and Van Rijn (1993). The first author gives a review of the classical formulations still in use today. Classical as well as original derivations are well documented by the second author.

One of the classical formulations quantifying the bed load transport is given by the Du Boys relation (Koutitas 1988):

$$q_b = \chi \tau_b (\tau_b - \tau_{cr}) \qquad (5.1)$$

where q_b refers to the bed load transport rate, τ_b is the bed shear stress, τ_{cr} is the critical bed shear stress above which sediments start to move, and χ is a (dimensional) coefficient dependent on the size, geometry and specific weight of the grains. Such a relation suggests the dependence of the bed load transport rate on the magnitude of the bed shear stress needed for the initiation of motion near the bed. Evidently, below a certain critical stress, no transport occurs. Several other relationships, which are still in use today, based on the influence of the bed-shear stress are reviewed in Koutitas (1988). These include the conventional Meyer-Peter and Kalinske-Frijlink formulae with some modifications introduced to allow interaction of current and waves on the effective bed shear stress.

The description of Van Rijn (1993) makes use of the saltation characteristics of the sediments derived using equations of particle motion. The bed load transport is then defined as a function of the particle velocity u_b, volumetric concentration c_b and bed load layer thickness δ_b as:

$$q_b = c_b u_b \delta_b \qquad (5.2)$$

The particle velocity is derived from a balance of forces acting on a particle which include gravity, drag and frictional forces. u_b is approximated by

$$u_b = u_* \left[9 + 2.6 \log(D_*) - 8 \left[\frac{\theta_{cr}}{\theta} \right]^{\frac{1}{2}} \right] \qquad (5.3)$$

where u_* is the bed shear velocity, D_* is a dimensionless particle parameter, θ and θ_{cr} the mobility and the critical mobility parameter respectively. The dimensionless particle parameter D_* is dependent on the particle diameter d:

$$D_* = d \left[\left[\frac{\rho_s}{\rho} - 1 \right] \frac{g}{v^2} \right]^{1/3} \qquad (5.4)$$

where ρ_s and ρ refer to the density of sediment particle and water respectively, g is the acceleration due to gravity and v is the kinematic viscosity coefficient of water. The bed load layer thickness is approximately equal to the saltation height of the particles. Van Rijn (1993) gives the approximate relation for δ_b in the case of an approximately flat bed:

$$\delta_b = 0.3 d_{50} D_*^{0.7} T^{0.5} \qquad\qquad (5.5)$$

where T is a dimensionless bed-shear stress parameter given by:

$$T = \frac{(\tau_b - \tau_{cr})}{\tau_{cr}} \qquad\qquad (5.6)$$

Flume experiments were conducted by Van Rijn (1993) to measure the bed load transport rate as a function of the particle characteristics. Using the measured bed load transport rates in conjunction with Equations (5.3 and 5.5), a function for the bed-load concentration is derived

$$c_b = 0.18 c_m \frac{T}{D_*} \qquad\qquad (5.7)$$

where c_b is the volumetric bed-load concentration and c_m is the maximum volumetric concentration. Using this value of c_b and approximate relationships for the saltation height and particle velocity, Van Rijn (1993) proposes several methods to estimate the bed-load transport rate. The following equations give a summary of these empirical relationships.

$$
\begin{aligned}
q_b &= 0.053(s-1)^{0.5} g^{0.5} d_{50}^{1.5} D_*^{-0.3} T^{2.1} &\quad if \quad& T < 3 \\
q_b &= 0.1(s-1)^{0.5} g^{0.5} d_{50}^{1.5} D_*^{-0.3} T^{1.5} &\quad if \quad& T \geq 3 \\
q_b &= 0.25 d_{50} g^{0.5} \frac{u}{C'} D_*^{-0.3} T^{1.5} & & \\
q_b &= 0.005 u h \left[\frac{u - u_{cr}}{\sqrt{(s-1) g d_{50}}} \right]^{2.4} \left[\frac{d_{50}}{h} \right]^{1.2} & & \\
q_b &= 0.00099 \rho_s [(s-1) g d_{50}]^{0.5} d D_*^{0.9} T^{2.4} & &
\end{aligned}
\qquad (5.8)
$$

Here d_{50} is the median particle diameter, u is the depth-averaged flow velocity, u_{cr} is the critical depth-averaged flow velocity, s is the sediment specific weights or relative density ($= \rho_s/\rho$), h is the water depth and C' is the grain-related Chezy-coefficient. All these formulae, except for the last, estimate the bed load transport rate from the general relation given by Equation (5.2). The last equation determines the bed load transport rate from the sediment pick-up rate and saltation length defined as $q_b = E \lambda_b$, in which E is described as the pick-up rate and λ_b is the saltation length (Van Rijn 1993).

A general observation in bed load transport researches supported by field measurements is the inverse effect of particle size to the estimated bed-load transport rate. The particle size determines the saltation characteristics of sediments and as such also determine the transport rate. Figure 5.1 shows the influence of the particle size on the bed-load transport rate.

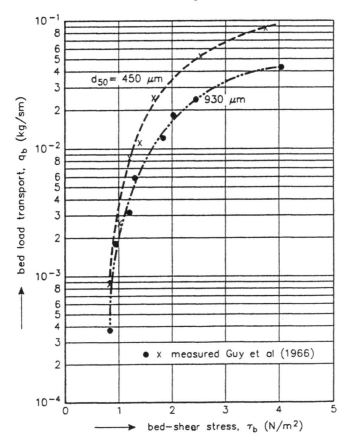

Figure 5.1. Influence of particle size on the bed-load transport rate (Van Rijn 1993).

Evidently, the bed-load transport rate is inversely proportional to the particle size. A decrease of the grain diameter by a factor of 2 yields an increase of the transport rate by a factor of 2 (Van Rijn 1993). This observation is likewise implicit in the empirical relationships for the bed load transport rates derived by Van Rijn (1993) since the bed shear stress parameter T gives correspondingly higher values for smaller particle sizes. Except for the fourth relation given in Equation (5.8), all of the other relations show the strong influence of the dimensionless bed-shear stress T on the bed load transport rate. The inherently inversely proportional relationship of T and the particle size (which outweighs the effect of the median particle diameter d_{50} in the given relations) also implies an inverse relation between the bed-load transport rate and the particle size.

Some approaches in determining the bed-load transport rate are based on stochastic methods. The approach given by Kalinske (1947), Einstein (1950) and Van Rijn (1987) estimates the transport rate from probabilistic functions related to the erodability of the particles on the sediment bed. Van Rijn (1993) gives a detailed description of these stochastic methods.

It should be noted that the sediment particles comprising the bed load may become suspended in the water column, i.e. fully supported by the fluid. This process is highly dependent on the influence of the bed-shear velocity and resulting shear stress acting above the bed load layer. In general, with increasing values of the bed shear velocity, the rolling and sliding particles at the bed load transport zone undergo saltations with intermittent contacts with the bed. When the value of the bed shear velocity exceeds the fall (or settling) velocity of the particles, the particles can be lifted to a level at which the upward turbulent forces become comparable with and eventually higher than the submerged weights of the particles with the result that the particles go into suspension (Van Rijn 1985). This later mode of sediment transport is discussed in the following section.

5.1.2. Suspended-Load Transport

Above the bed-load transport zone, particles remain suspended in the water column by way of the hydrodynamic lift and buoyancy forces exceeding the submerged weights and inter-particle forces of attraction of the sediment particles. The movement of such particles presumably moving with the fluid velocity is known as the suspended load transport. The suspended load is the sediment volume moving according to this mode of motion and its magnitude depends on the ratio of the bed shear velocity and the particle fall velocity (Koutitas 1988). Generally, the bed shear velocity must exceed the particle fall velocity to keep the sediments in suspension. If the bed shear velocity and the particle fall velocity are represented by u_* and w_s respectively, then the criterion for particle suspension according to Bagnold is $u_* > w_s$ (Van Rijn 1984).

The concentration of suspended particles is generally very high near the bed, and decreases rapidly towards the water surface. A hypothetical distribution of suspended sediment concentration in the vertical is shown in Figure (5.2). A reference level a is assumed, separating the bed load from the suspended load transport layers. Assuming a logarithmic velocity profile and this sediment concentration profile, the suspended load transport can be highly non-linear in the vertical. Above the bed load layer where the concentration is high and the horizontal fluid velocity has a considerable magnitude, the suspended load transport reaches a maximum (Smith 1992, Van Rijn 1993).

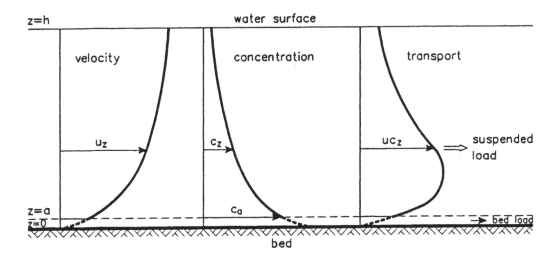

Figure 5.2. Suspended sediment transport profile according to Van Rijn (1993).

The actual suspended sediment profile in the coastal zone is three-dimensional in nature. Assuming that the suspended sediments does not influence the hydrodynamics of the flow, and that the fall (settling) velocity of the suspended particles w_s can be assumed constant, then the field equation describing the time-dependent mass conservation of the suspended sediments in nearly horizontal geophysical flows is given by;

$$\frac{\partial c}{\partial t} + u\frac{\partial c}{\partial x} + v\frac{\partial c}{\partial y} - w_s\frac{\partial c}{\partial z} = \frac{\partial}{\partial x}\left(K_x\frac{\partial c}{\partial x}\right) + \frac{\partial}{\partial y}\left(K_y\frac{\partial c}{\partial y}\right) + \frac{\partial}{\partial z}\left(K_z\frac{\partial c}{\partial z}\right) \qquad (5.9)$$

where c is the suspended sediment concentration and K_x, K_y, and K_z are the diffusion coefficients in the x, y and z directions respectively. This equation describes in full the concentration profile of suspended matter in a turbulent advective flow field (Koutitas (1988).

The existence of vertical concentration gradients of suspended matter in the water column depends on the ratio of turbulent dispersion and sedimentation (Lijklema et al. 1994). The characteristic time scales for dispersion t_d and settling t_s can be compared as in

$$\frac{t_d}{t_s} = \frac{h^2/2K_z}{h/w_s} = \frac{w_s h}{2K_z} \qquad (5.10)$$

where K_z is the vertical dispersion coefficient. Accordingly, if $t_d/t_s << 1$ then dispersion will be dominating. In many surface water systems such as shallow lakes, it can be shown that dispersion dominates over sedimentation. This means that concentration gradients in the vertical can be assumed negligible in these cases. Hence, the mass conservation equation represented in Equation (5.9) can be reduced to a two-dimensional horizontal equation, i.e. the vertical concentration gradients can be eliminated.

The sediment concentration profile in the vertical direction can be determined by assuming a uniform flow and sediment equilibrium conditions such that the upward flux of sediment particles is balanced by the downward flux due to settling, i.e.

$$K_z \frac{\partial c}{\partial z} + w_s c = 0 \qquad (5.11)$$

The solution of this equation needs an approximation of the vertical eddy diffusion (dispersion or mixing) coefficient. Several distribution functions of this variable are described in literature, i.e. a constant distribution, linear, parabolic, and a parabolic-constant (see Figure 5.3). Assuming a linear shear stress distribution from the reference level a to the water surface, a parabolic distribution function for the diffusion coefficient is obtained, i.e.

$$K_z = \beta \kappa u_* \frac{z(h-z)}{h} \qquad (5.12)$$

where κ is the Von Karman constant ($= 0.4$) and β is a constant of proportionality between the diffusion coefficients for suspended sediment and fluid mass (Dyer 1986). Using a reference concentration c_a at the reference level a as a bottom boundary condition, integration of Equation (5.11) in conjunction with Equation (5.12) results in the classical Rouse profile for the sediment concentration given by

$$c_z = c_a \left[\frac{h - z}{z} \frac{a}{h - a} \right]^{z_*} \qquad (5.13)$$

where z_* is the Rouse or suspension number which expresses the influence of the upward turbulent fluid forces and the downward gravitational forces (Van Rijn 1984b). This is given by

$$z_* = \frac{w_s}{\beta \kappa u_*} \qquad (5.14)$$

in which u_* ($= \tau_b/\rho$) is the bed-shear (friction) velocity. It should be noted that, analogous to Equation 5.10, the z_* parameter is of primary importance in describing the suspended sediment distribution over the water depth. For low values of z_* ($z_* < < 1$), the suspended sediments are almost uniformly distributed over the water depth. For z_* values larger than unity, the suspended sediments are confined in the near-bed layer upto mid-depth.

The Rouse concentration profile (Equation 5.13) gives a zero sediment concentration at the water surface which is less realistic. To obtain a non-zero concentration at the surface, Van Rijn (1993) assumed a parabolic-constant distribution of the eddy diffusion coefficient. For the lower half of the flow depth (bed-load transport zone until mid-water depth), the parabolic distribution of eddy diffusion (Equation 5.12) is assumed. For the upper half of the flow depth (mid-depth to the water surface), the eddy diffusion function is assumed constant, i.e.

$$K_z = \frac{1}{4} \beta \kappa u_* h \qquad for \qquad \frac{z}{h} \geq 0.5 \qquad (5.15)$$

The resulting sediment concentration profile for the lower half of the water column, as described by the parabolic eddy distribution, is given by the Rouse equation (Equation 5.13). Above this zone with a constant eddy diffusion coefficient, the sediment concentration profile is given by

$$c_z = c_a \left(\frac{a}{h-a} \right)^{z_*} e^{-4z_*(z/h - 0.5)} \qquad for \qquad \frac{z}{h} \geq 0.5 \qquad (5.16)$$

The vertical distribution profiles of the sediment concentration described by various profiles for the fluid and sediment diffusion (mixing) coefficients are illustrated in Figure (5.3). The sediment and fluid diffusion coefficients are related by $K_z = \beta \phi \epsilon_f$, where ϵ_f is the fluid diffusion coefficient, and ϕ is a coefficient expressing the effect of the suspended sediments on the turbulence structure of the fluid (Van Rijn 1993).

The sediment concentration profile expressed by Equation (5.16), in conjunction with the Rouse concentration profile (Equation 5.13), generally gives good agreement between observed and measured suspended sediment concentrations (Van Rijn 1993).

Assuming a logarithmic velocity profile and the sediment concentration profiles represented by Equations (5.13 and 5.16), the depth-averaged suspended load transport rate q_s can be estimated from (Koutitas 1988);

$$q_s = \frac{\left[\left(\frac{a}{h} \right)^{z_*} - \left(\frac{a}{h} \right)^{1.2} \right] u h c_a}{\left(1 - \frac{a}{h} \right)^{z_*} (1.2 - z_*)} \qquad (5.17)$$

where u is the depth-averaged flow velocity. The suspension number z_* includes the effect of both current and waves, such that with $\beta = 1$, this is given by (Koutitas 1988);

$$z_* = \frac{w_s}{\kappa} \left[\frac{\tau_{cw}}{\rho} \right]^{-1/2} \qquad (5.18)$$

Here, τ_{cw} is the effective bed shear stress due to current and waves. The bed shear stress is assumed to be a non-linear product of the current-related stress ($\tau_c = \rho g u^2 / C^2$) and the peak orbital velocity averaged during a wave cycle. An empirical relation quantifying the total stress due to current and waves according to Bijker's analysis is given by (Dyer 1986, Koutitas 1988):

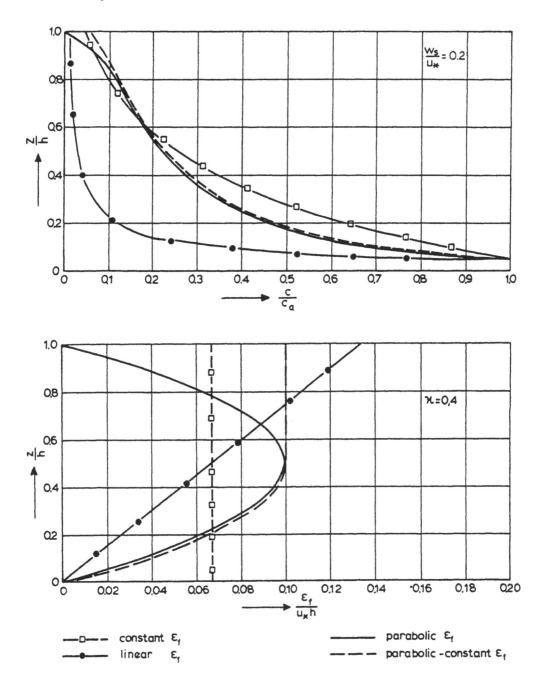

Figure 5.3. Concentration profiles described by the various assumptions of fluid and sediment diffusion coefficients after Van Rijn (1993). ϵ_f is the fluid mixing coefficient.

$$\tau_{cw} = \tau_c \left[1 + \frac{1}{2} \left[\xi \frac{u_m}{u} \right]^2 \right] \tag{5.19}$$

where u_m refers to the near-bed wave orbital velocity and ξ is a dimensionless parameter related to the bed roughness as in

$$\xi = \frac{C\sqrt{f_w}}{\sqrt{2g}} \tag{5.20}$$

in which f_w is the bed friction coefficient for waves, and C is the Chezy coefficient. Estimation of the suspended sediment transport rate using Equation (5.17) usually gives reasonable values for z_* values between 0.1 and 3 (Koutitas 1988).

Several mathematical equations for the suspended load transport can be found in literature. These include the formulae proposed by Einstein (1950), Bagnold (1966), Bijker (1971) and Van Rijn (1984b). These are reviewed by Van Rijn (1993). When the vertical profiles of flow velocity and sediment concentration are known, the suspended sediment load can be obtained by integration of the relations:

$$q_{sx} = \int_{-h+a}^{0} u c \, dz \quad , \quad q_{sy} = \int_{-h+a}^{0} v c \, dz \tag{5.21}$$

where q_{sx}, q_{sy} are the volumetric suspended load transport rates, u and v refer to the current velocities in the x and y-directions respectively and h is the effective water depth (Koutitas 1988). Note that the integration is performed not from the bed, but at the reference level a above the bed-load transport zone.

It should be noted that the suspended (and bed) load transport rates obtained in the preceding relationships use the volumetric concentration. The weight concentration is obtained by multiplying the volumetric concentration by the sediment density ρ_s.

5.2. Resuspension and Sedimentation

Most of the sediment particles in the water column which constitute the suspended load participate in a continuous process of resuspension and sedimentation. Particles resting on the sediment bed are resuspended if the submerged weight and attractive forces are overcome by the hydrodynamic lift and buoyancy forces. Such overcoming forces are characteristic for the fluid in which the particles are submerged. Of particular importance is the turbulence intensity of the water column related to the velocity fluctuations near the bottom. The effective turbulence is related to the wave and flow-induced motion near the bed resulting in a specific shear stress capable of eroding the bottom sediments if it exceeds the critical stress for sediment resuspension. This critical stress is characteristic of the submerged particles but may be modified and dictated in part by the fluid physico-chemical characteristics. The following sections give a description of the sediment fluxes due to resuspension and sedimentation, taking into account the important factors affecting both processes.

5.2.1. Resuspension by Waves and Currents

Resuspension is taken to be related to both wave and current-induced bottom stress. In some research studies especially applied to shallow lakes, wind-induced waves are known to dominate the resuspension process such that the effect of the current is neglected. Studies conducted in Lake Balaton (Hungary) by Luettich (1987), and in Lake Veluwe and Lake Marken (The Netherlands) by Blom et al. (1992) and Van Duin (1992) all showed the dominant effect of waves in inducing resuspension. On the other hand, it is well established that the resuspension process and gross sediment transport processes in rivers depend heavily on the flow characteristics of the water. In the coastal zone, both effects are deemed important. While waves may dominate sediment resuspension at a given time, the current acts to complement in a nonlinear way the resuspension and total sediment transport. But in cases when these wind-generated waves are very weak, it can be assumed that sediment resuspension and transportation are governed by the current, tide or wind-induced. Nevertheless the wind is almost ceaseless and it is essentially the interaction of both wind-generated waves and the tidal (and wind-driven) current that makes transport processes more complex in the coastal sea. As pointed out by Koutitas (1988), the waves act as a destabilization, mobilization and suspension factor for the sediments in the coastal zone, and a minimal current (even in the form of Stokes' wave drift), may be able to carry away the already activated sediment grains. In essence, the effect of both current and waves in the sediment resuspension and transport process can be viewed as inseparable. The following sections seek to give a description on the resuspension and sedimentation processes in the coastal sea taking into consideration the independent and combined effect of waves and current.

5.2.1.1. Wave-Induced Resuspension

Wave-induced resuspension takes into account the characteristics of the wind-generated surface waves. The significant wave characteristics defined in the previous chapter gives an empirical estimation of the significant wave height, period and length from known wind-forcing and basin characteristics such as depth and geometry (fetch). These surface waves riding atop the tide and wind-generated surface elevation, have a period scaled in seconds, and wavelengths scaled in meters. As a consequence, the associated wave-boundary layer does not have a chance to grow to a thickness of more than a few millimeters (Luettich et al. 1990). Since bottom shear is proportional to the velocity gradient in the boundary layer, the bed shear stress due to waves can be very significant when considering resuspension of bottom sediments.

The orbital velocity u_m induced by the surface waves at the sediment-water interface can be derived from the wave characteristics. The maximum value of u_m near the sediment bed is given by (Phillips 1966):

$$u_m = \frac{\pi H_s}{T_s \sinh\left[\dfrac{2\pi h}{L_s}\right]} \tag{5.22}$$

where H_s, T_s and L_s are the significant wave height, period and length respectively. The estimated near-bed orbital velocity is then used to calculate the bed-shear stress τ_w due to the waves. Several approximations of the bed-shear stress are available in literature. They have in common that the stress is proportional to the square of the maximum wave orbital velocity. Van Rijn (1993) assumes

$$\tau_w = \frac{1}{4}\rho f_w (u_m)^2 \tag{5.23}$$

where ρ is the water density, and f_w is a wave friction factor which is assumed to be related to the near-bed wave orbital excursion A and the wave-related bed roughness height $k_{s,w}$ as in (Van Rijn 1993):

$$f_w = \exp\left[-6 + 5.2\left[\frac{A}{k_{s,w}}\right]^{-0.19}\right]$$ (5.24)

The wave orbital excursion near the bed is similarly estimated from the significant wave characteristics and is given by;

$$A = \frac{H_s}{2\sinh(2\pi h/L_s)}$$ (5.25)

The wave-related roughness height $k_{s,w}$ may range from 0.01 to 0.1 m (Van Rijn, 1993).

The resuspension flux of bottom sediments by the wave-induced characteristics is basically the upward transport of sediment mass brought into suspension per unit area and unit time. Several empirical relationships for this flux are available in literature. Aalderink et al. (1984), Van Duin (1992) and Lijklema et al. (1994) give a review of the available methods for estimating the sediment resuspension flux. In Aalderink et al. (1984), the wave-induced resuspension flux is estimated from an energy balance. Accordingly, the potential energy attributed to the resuspension flux is approximately equal to the energy dissipated at the bottom. Using available relationships for the terms in the energy balance equation, the following empirical relationship for the wave-induced resuspension flux is obtained:

$$\phi_r = kn^{1.125}u_m^{1.875}$$ (5.26)

where k is a constant dependent on several factors including sediment density and n is the wave frequency.

The empirical model according to Luettich (1987) relates the resuspension flux directly to the significant wave height by:

$$\phi_r = K \left[\frac{H_s - H_{cr}}{H_{ref}} \right] \tag{5.27}$$

where H_s, H_{cr}, H_{ref} are the significant wave height, critical wave height and reference height respectively.

Van Duin (1992) uses the relationship given by Lam and Jacquet (1979) and expresses the resuspension flux as dependent on the wave orbital velocity and a critical value of this orbital velocity as in:

$$\phi_r = k(u_m - u_{mc}) \tag{5.28}$$

where u_{mc} is the critical wave orbital velocity, and k is the resuspension constant given by:

$$k = \frac{K\rho_s\rho}{(\rho_s - \rho)u_m} \tag{5.29}$$

in which K is a constant. With appropriate empirical constants determined from calibration, most of the foregoing flux formulations are known to give reasonable estimates of the wave-induced resuspension flux of bottom sediments in shallow lakes. For example, Blom et al. (1992) used Equation (5.28) to estimate the resuspension flux in the STRESS-2D model developed by the authors and found good agreement between the measured and simulated suspended sediment concentrations in some shallow lakes in the Netherlands.

In coastal and estuarine studies, stress-based resuspension flux formulations are applied. The relationship proposed by Parchure and Mehta (1985) treats resuspension of homogeneous or non-homogeneous (cohesive) sediment beds. Accordingly, the flux of resuspension for homogeneous densely consolidated bed can be described by the commonly-used Partheniades erosion function

$$\phi_r = M \left[\frac{\tau_b - \tau_{cr}}{\tau_{cr}} \right] \tag{5.30}$$

where M is an erosion parameter, τ_b and τ_{cr} refers to the wave-induced bed shear stress and critical shear stress respectively. The value of M and τ_{cr} depends on certain physico-chemical parameters such as the cation exchange capacity (CEC) characterizing the interparticle bond strength, Sodium Adsorption Ratio (SAR) and pH (Parchure and Mehta 1985). In cases where the bed is non-homogeneous, soft and partly consolidated, the flux of resuspension is found to be;

$$\phi_r = k_r \exp[\alpha(\tau_b - \tau_{cr})^{1/2}] \qquad (5.31)$$

where k_r is regarded as the floc erosion rate, and α is a parameter which is known to be inversely proportional to the absolute temperature. This formulation implies that the resuspension flux has the value k_r if τ_b does not exceed the critical stress τ_{cr}. Parchure and Mehta (1985) pointed out that as a consequence of the stochastic nature of τ_b some entrainment of sediment flocs will occur even when $\tau_b \leq \tau_{cr}$.

5.2.1.2. Flow-Induced Resuspension

Under certain circumstances, the flow velocity can be assumed to dominate the resuspension process of particulate material. In fact, the magnitude of this flow velocity or the concomitant bed stress, is the basis of erosion and transportation formulations for river systems.

In the coastal zone, resuspension by the tide-induced flow velocities can be discerned especially when surface waves are weak (as when the wind is weak for instance). The characteristic wavelength and period of the tide and wind-generated long wave currents responsible for the subsequent flows are in the order of several kilometers and hours respectively. The current boundary layer has a characteristic period on the order of hours and therefore could grow to a thickness comparable to the water depth (Luettich et al. 1990). As a consequence, the associated bottom stress due to the mean flow is generally less than the bottom stress generated by the short (surface) waves. In the absence of surface waves thus the resuspension must be caused by the long-wave tidal currents provided these are strong enough. Several studies confirm the strong influence of tidal currents on sediment resuspension. For example, Pejrup (1988) studied the suspended sediment dynamics in the Danish part of the Wadden Sea and found that erosion-transportation cycles towards the land and back to the sea are dependent on the tidal cycle which shows that tide-induced flow velocities are responsible for the resuspension of particulate materials in whichever fashion they may be transported. In general there is an asymmetrical transport process due mainly to the asymmetry in the flood and ebb currents. The same phenomenon is noted by Van Leussen (1994) in his study at the Ems Estuary (The Netherlands). Observations of the dynamics of suspended macroflocs showed that the flood and ebb currents give rise to an asymmetric stepwise resuspension and transport of sediments in the estuary. Higher concentrations of suspended materials during increasing tide currents are observed confirming the dependence of resuspension on the strength of the flow velocities.

As the wind is ceaseless, the actual resuspension and transportation processes in the coastal zone can be quite complex due to the presence of both wind-generated and tide-induced flow velocities interacting non-linearly. The individual effects of the wind and the tide cannot, in a strict sense, be decoupled. Field studies could only measure the total magnitude of the flow velocity and relate it to the observed dynamics of suspended sediment concentrations. In such studies, the magnitude of the bed shear stress estimated from the combined wind and tide-induced current can be related to the resuspension flux of particulate materials. The equations (5.29-5.30) defining the flux of resuspension as described by Parchure and Mehta (1985) and Mehta et al. (1989) are applicable in such instances. However, the bed shear stress is now dependent on the magnitude of the effective current velocity. The current-related bed shear stress can be estimated from (Van Rijn 1993);

$$\tau_c = \frac{1}{8}\rho f_c V^2 \tag{5.32}$$

in which $V (= [u^2 + v^2]^{1/2})$ is the magnitude of the flow velocity and f_c is the current-related friction coefficient, assumed to be a function of the effective water depth h and the current-related roughness height $k_{s,c}$ as in:

$$f_c = \frac{0.24}{\log^2(12h/k_{s,c})} \tag{5.33}$$

The magnitude of the bed stress is then substituted in the expressions of Parchure and Mehta (1985) given by Equations (5.30-5.31) to obtain the flow-induced resuspension flux. It should be noted that the resuspension parameters M and k_r depend on the erodability of the sediment bed and hence are site-specific. The critical shear stress τ_{cr} is also dependent on the type and composition of sediment particles on the bed. In natural systems, where the bed is composed of a mixture of sediments varying in sizes and compositions, this critical shear stress has no unique value. Wilcock (1993), however, made a study on the critical shear stress of natural sediments and found it to be approximately in the range given by the conventional Shields approach. Van Rijn (1993) gives an analytical relationship for the critical shear stress according to Shields as

$$\tau_{cr} = (\rho_s - \rho) g d_{50} \theta_{cr} \qquad (5.34)$$

where θ_{cr} is defined as the critical particle mobility parameter. The value of θ_{cr} varies for different particle sizes and densities, i.e.

$$
\begin{aligned}
\theta_{cr} &= 0.24 D_*^{-1} & 1 &< D_* \leq 4 \\
\theta_{cr} &= 0.14 D_*^{-0.64} & 4 &< D_* \leq 10 \\
\theta_{cr} &= 0.040 D_*^{-0.1} & 10 &< D_* \leq 20 \\
\theta_{cr} &= 0.013 D_*^{0.29} & 20 &< D_* \leq 150 \\
\theta_{cr} &= 0.055 & D_* &> 150
\end{aligned}
\qquad (5.35)
$$

in which D_* is the particle parameter given in Equation (5.4). The median particle diameter d_{50} is usually taken to characterize the particle parameter. The Shields approach can be very useful in estimating the critical shear stress of generally non-cohesive particles with particle sizes above the silt-clay range. Reported values for the cohesive fractions from laboratory experiments vary strongly, which indicates that the critical shear stress could be very site-specific.

5.2.2. Sedimentation

Sedimentation is usually taken to mean gravitational settling. The sedimentation flux ϕ_s is defined as the downward transport of sediment mass per unit area per unit time. The usual expression quantifying the sedimentation flux in surface water systems is related to the sediment settling velocity w_s and near-bed concentration c_b as in

$$\phi_s = w_s c_b \qquad (5.36)$$

The sedimentation flux represented by this equation is taken just above the bed-load layer, i.e. at the reference level a. Quantification of the sedimentation flux from the foregoing relation is not as straightforward as it seems. The estimation of the settling velocity, especially for fine cohesive sediment fractions, and the reference concentration are complicated by several factors related to the sediment particles in suspension and the turbulence structure of the water column.

Krone (1962) proposed that the sedimentation flux should be derived using the probability that the suspended sediments will be deposited. The probability of sediment deposition p is assumed to be a function of the shear stress near the bed, i.e. $p = (1 - \tau_b/\tau_{cd})$ where τ_b and τ_{cd} are the near-bed shear stress and critical shear stress for deposition respectively. Accordingly, $p = 0$ when $\tau_b \geq \tau_{cd}$. The modified sedimentation flux as proposed by Krone (1962) is then estimated from

$$\phi_s = p w_s c_b \tag{5.37}$$

The estimation of near-bed concentration c_b is an important issue in sediment transport studies. It has been pointed out by Dyer (1986) that its quantification is one of the weakest points of sediment transport modelling. The time-dependent sediment concentration in the water column is controlled by the local sedimentation flux at the reference level (above the bed-load transport zone) whose mean concentration can be assumed equal to c_b. It should be noted that the expression (Equation 5.7) derived by Van Rijn (1993), yields a zero-near bed concentration if the bed-shear stress is less than the critical shear stress, which is unrealistic. Mehta et al. (1989) assumed that the near-bed concentration is related to the depth-averaged sediment concentration and to the vertical variation in the concentration. The analytic expression quantifying this relation is due to Teeter (1986) and is given by;

$$c_b = c \left[1 + \frac{P_e}{1.25 + 4.75 p^{2.5}} \right] \tag{5.38}$$

where c is the depth mean concentration, p is the probability of deposition, and P_e is the Peclet number (analogous to the suspension number z_*) given by

$$P_e = \frac{w_s h}{K_z} = \frac{w_s}{\frac{1}{6} \beta \kappa u_*} \tag{5.39}$$

in which K_z is the depth-averaged eddy diffusivity, whose approximate value has been obtained by vertical integration throughout the water depth of the parabolic distribution given by Equation (5.12), u_* is the friction velocity (τ_b/ρ) derived from both current and waves, and κ is the von Karman constant. Normally, β can be assumed equal to 1 (Koutitas 1988). In general, an increasing magnitude of P_e implies an increasingly deposition-dominated or net sedimentation environment, which is the case in many estuaries (Mehta et al. 1989).

When vertical gradients in the concentration are negligible, the sedimentation flux ϕ_s can be assumed equal to $w_s c$ or $w_s(c - c_o)$ where c_o is taken as a background concentration for very slow settling or non-settling materials in the water column. This later approximation is usually applied in shallow lakes (see for example Blom and Toet 1991, Lijklema et al. 1994).

As a first approximation, the settling velocity can be estimated from the Stokes settling rate theory. For homogeneous spherical particles settling in an undisturbed water column w_s is given by

$$w_s = \frac{1}{18}\left[\frac{\rho_s - \rho}{\rho}\right]\frac{g\,d_s^2}{\nu\Theta} \tag{5.40}$$

where d_s is the diameter of a spherical particle, ν is the kinematic viscosity of water and Θ is a coefficient of form resistance. For particles of various sizes, approximate relations for the determination of the settling velocity are given by (Koutitas 1988);

$$w_s = \frac{1}{18}\frac{(s - 1)g\,d_{50}^2}{\nu}\,, \qquad\qquad d_{50} < 100\mu$$

$$w_s = \frac{10\nu}{d_{50}}\left[\left[1 + \frac{0.01(s - 1)g\,d_{50}^3}{\nu^2}\right]^{0.5} - 1\right], \quad 100\mu < d_{50} < 1000\mu \tag{5.41}$$

$$w_s = 1.1[(s-1)g\,d_{50}]^{0.5}\,, \qquad\qquad d_{50} > 1000\mu$$

where s is the specific weight (or relative density) of the sediment particles. Estimation of the settling velocity in natural suspensions is complicated not only by the heterogeneity of the particles in the water column, but also by the turbulence in the water column. Suspended particles of varying sizes, density and composition, are in a constant motion from high to low (or vice versa) zones of turbulence intensity, so that the validity of the Stokes settling rate theory is questionable due to the turbulence.

When sediment concentrations are high (several thousands mg/l), Van Rijn (1993) suggests that the settling velocity is not constant but depends on the concentrations because hindered settling occurs. The relationship suggested to represent this has the same form as in Mehta (1986):

$$w_{s,m} = (1 - c)^{\gamma} w_s \qquad (5.42)$$

where $w_{s,m}$ is the particle settling velocity in a mixture, w_s is the particle settling velocity in clear and still fluid, and γ is a coefficient (4 to 5) for particles of diameter 50 to 500 μ.

Van Leussen (1994), who studied intensively the characteristics of flocculated cohesive sediments, proposed a different method in estimating the particle settling velocity. From a number of observations with field settling tubes, a general exponential relation between the settling velocity and the sediment concentration was observed. Hence, he proposed the following power relation:

$$w_s = K C^m \qquad (5.43)$$

where K and m are empirical constants which depend on the turbulence structure of the flow, the cohesivity of the suspended particles, temperature, and the concentrations of several dissolved and colloidal constituents in the water column (Van Leussen 1994).

5.3.   Modelling Suspended Sediment Transport in the Coastal Sea

Sediment transport modelling as addressed in this study focuses on the problem of suspended sediments. More attention is needed in this respect than to the bed-load transport process since one of the main problems addressed in this thesis is the reduced light penetration as affected by suspended sediments in the water column.

Modelling of suspended sediments had been the subject of many researchers, e.g. Wang (1989), Lee et al. (1994) and other authors. Mostly, the modelling approach is based on the first or second-order mass-conservation laws solved using finite difference or finite element methods. The modelling study presented here discusses a third-order method based on the explicit finite difference technique proposed by Ekebjærg and Justesen (1991). The method is used for reason of its simplicity, efficiency and compatibility with the hydrodynamic model presented in the previous chapter. The general numerical scheme (called QUICKEST) is extended to include the source-sink terms to account for the effects of resuspension and sedimentation on the dynamics of suspended sediments in the coastal zone. The following sections discuss the modelling approach used in this study.

5.3.1. Third-Order Suspended Sediment Transport Model

The suspended sediment transport model used in this study is based on the numerical procedure described by Ekebjærg and Justesen (1991). The governing equation is basically obtained by vertical integration of Equation (5.9). This gives,

$$\frac{\partial c}{\partial t} + u \frac{\partial c}{\partial x} + v \frac{\partial c}{\partial y} = K_x \frac{\partial^2 c}{\partial x^2} + K_y \frac{\partial^2 c}{\partial y^2} + \frac{\phi_r - \phi_s}{h} \qquad (5.44)$$

where c is the depth-averaged concentration of suspended sediments, u and v represent the x and y-components of the depth mean current velocities determined from the hydrodynamic model, K_x and K_y are (depth-mean) dispersion coefficients, ϕ_r is the resuspension flux taken as a source term, ϕ_s is the sedimentation flux (sink term), and h is the effective water depth ($= h_o + \zeta$). The first term on the left of Equation (5.44) represents the local change of suspended sediment concentration. The next two terms described the changes in sediment concentrations due to advection by currents. The effect of dispersion (or diffusion) is represented by the first two terms on the right followed by the source and sink terms, i.e. the fluxes of resuspension and sedimentation.

The finite difference approximation of Equation (5.44) can be obtained using Taylor series expansion around the point i,j (Ekebjærg and Justesen 1991). Disregarding the source-sink terms, this gives

$$\frac{c_{i,j}^{n+1} - c_{i,j}^{n}}{\Delta t} + u_{i,j}^{n} \frac{c_{i+1,j}^{n} - c_{i-1,j}^{n}}{2\Delta x} + v_{i,j}^{n} \frac{c_{i,j+1}^{n} - c_{i,j-1}^{n}}{2\Delta y}$$
$$= K_x \frac{c_{i+1,j}^{n} - 2c_{i,j}^{n} + c_{i-1,j}^{n}}{\Delta x^2} + K_y \frac{c_{i,j+1}^{n} - 2c_{i,j}^{n} + c_{i,j-1}^{n}}{\Delta y^2} + TE \qquad (5.45)$$

TE represents higher order truncation error terms which result from approximating the partial derivatives of Equation (5.44) by finite differences. Recent researches in advection-diffusion modelling showed that these truncation terms should be included to obtain more accurate predictions and attain mass conservation. There are several mass-conserving schemes available in literature (see for example Leonard (1991), Zoppou and Roberts (1993) and Li and Yu (1994)). The numerical procedures proposed in these papers included either second or third order truncation terms to obtain higher accuracy in their solution. Recently, Le Veque (1996) also presents a series of mass-conserving numerical schemes involving the use of shape-preserving interpolating functions. The QUICKEST numerical scheme which is adopted in this study is also shown by Ekebjærg and Justesen (1991) to be mass conservative. In several test cases involving steady flow field, the scheme produces very little mass falsification. Following Ekebjærg and Justesen (1991), third order accuracy in space and time can be obtained by including the truncation error terms in Equation (5.45) as

$$TE = \frac{\Delta t}{2}\frac{\partial^2 c}{\partial t^2} + \frac{\Delta t^2}{6}\frac{\partial^3 c}{\partial t^3} + u\frac{\Delta x^2}{6}\frac{\partial^3 c}{\partial x^3} + v\frac{\Delta y^2}{6}\frac{\partial^3 c}{\partial y^3} + ... \qquad (5.46)$$

where the dots indicate fourth and higher order terms which are neglected by the authors. The time derivatives of concentration in this equation can be eliminated by differentiating Equation (5.44) with respect to time neglecting the source-sink terms (i.e. the resuspension and sedimentation terms). This yields,

$$\frac{\partial^2 c}{\partial t^2} = u^2\frac{\partial^2 c}{\partial x^2} + 2uv\frac{\partial^2 c}{\partial x\partial y} + v^2\frac{\partial^2 c}{\partial y^2} - 2uK_x\frac{\partial^3 c}{\partial x^3} - 2vK_x\frac{\partial^3 c}{\partial x^2\partial y}$$
$$- 2uK_y\frac{\partial^3 c}{\partial x\partial y^2} - 2vK_y\frac{\partial^3 c}{\partial y^3} \qquad (5.47)$$

$$\frac{\partial^3 c}{\partial t^3} = -u^3\frac{\partial^3 c}{\partial x^3} - 3u^2 v\frac{\partial^3 c}{\partial x^2\partial y} - 3uv^2\frac{\partial^3 c}{\partial x\partial y^2} - v^3\frac{\partial^3 c}{\partial y^3} \qquad (5.48)$$

for the second and third derivatives respectively. Substituting these equations in Equations (5.46 and 5.45) yields a third-order partial differential equation of the form

$$\frac{\partial c}{\partial t} = -u\frac{\partial c}{\partial x} - v\frac{\partial c}{\partial y} + K_x\frac{\partial^2 c}{\partial x^2} + K_y\frac{\partial^2 c}{\partial y^2} + \frac{\Delta t}{2}\left[u^2\frac{\partial^2 c}{\partial x^2}\right.$$

$$+ 2uv\frac{\partial^2 c}{\partial x\partial y} + v^2\frac{\partial^2 c}{\partial y^2} - 2uK_x\frac{\partial^3 c}{\partial x^3} - 2vK_x\frac{\partial^3}{\partial x^2\partial y}$$

$$- 2uK_y\frac{\partial^3 c}{\partial x\partial y^2} - 2vK_y\frac{\partial^3 c}{\partial y^3}\left.\right] + \frac{\Delta t^2}{6}\left[-u^3\frac{\partial^3 c}{\partial x^3} - 3u^2 v\frac{\partial^3 c}{\partial x^2\partial y}\right.$$

$$- 3uv^2\frac{\partial^3 c}{\partial x\partial y^2} - v^3\frac{\partial^3 c}{\partial y^3}\left.\right] + u\frac{\Delta x^2}{6}\frac{\partial^3 c}{\partial x^3} + v\frac{\Delta y^2}{6}\frac{\partial^3 c}{\partial y^3} + \frac{\phi_r - \phi_s}{h} \qquad (5.49)$$

Equation (5.49) is the basic equation for the suspended sediment transport model applied in this study. Its solution and application are described in Section (5.3.3).

5.3.2. Description of Resuspension and Sedimentation Fluxes

The resuspension flux is modelled using the expression (Equation 5.32) obtained by Parchure and Mehta et al. (1985). The bed stress in this study is assumed to depend on both current and waves. The total shear stress τ_{cw} can be assumed as an additive function of the stress due to current and that due to wave (Van Rijn, 1993). The current-related shear stress can be estimated from the depth-mean currents (Equation 5.32) provided by the hydrodynamic model. The depth-mean currents are assumed to be both wind and tide-driven as the hydrodynamic model includes both driving forces. On the other hand, the wave-induced bed shear stress can be estimated from the maximum wave orbital velocity attributed to the wind-induced waves. The characteristics of these surface waves were discussed in the previous chapter. The wave orbital velocity is estimated using Equation (5.22) and the wave-induced shear stress is likewise estimated using Equation (5.23). The total bottom stress for the resuspension flux is thus given by

$$\tau_{cw} = \frac{1}{8}\rho f_c[u^2 + v^2] + \frac{1}{4}\rho f_w(u_b)^2 \qquad (5.50)$$

where f_w and f_c are the wave and current-related friction factors respectively, which are estimated according to Equations (5.24 and 5.33). Equation (5.50) does not account for non-linear interaction between wave and current. As in Lee et al. (1994), the wave-related stress is a mean stress, integrated during a wave period.

On the other hand, the sedimentation flux is estimated from the relationship described by (Blom and Toet 1991, Lijklema et al. 1994), i.e. $\phi_s = w_s(c - c_o)$. A mean settling velocity, representative of a particular sediment fraction, is used in this study.

5.3.3. Numerical Solution by Explicit Finite Difference Technique

Similar to the hydrodynamic models, the numerical solution of the suspended sediment transport model described in this study is based on explicit finite difference technique. The use of such technique is favorable due to its simplicity and compatibility with the hydrodynamic model described in the previous chapter. This extended model makes use of the QUICKEST numerical scheme described by Ekebjærg and Justesen (1991). The scheme is essentially based on the one-dimensional upstream model of Leonard (1979) and extended in two dimensions by the authors. In the present application, the space staggered grid (Figure 4.1) described in the previous chapter is used wherein the sediment concentrations are situated at the center of a grid cell together with the water depth and surface elevation. The space derivatives in Equation (5.49) are approximated by upstream finite differences as in Ekebjærg and Justesen (1991). The resuspension and sedimentation fluxes are solved on the center of the grid cells. For numerical details, please see Appendix 2.

5.3.4. Open Boundary Condition

The open boundary condition for the sediment transport model developed in this study is essentially based on the Orlanski Radiation Condition proposed by Chapman (1985) for circulation modelling. Discussion of the concept was provided in the previous chapter (see Section 4.3.2). Its application in sediment transport modelling is discussed in this section.

The sediment concentrations, being scalar quantities, are treated at the open boundary using the same concept of propagation (or radiation) as in the hydrodynamic model. Ideal solutions for the sediment concentrations at the open boundaries must not indicate any reflection of advected concentrations. In that case, the predicted concentrations inside the modelled area are not corrupted by computational errors introduced at the boundaries. The use of several open boundary conditions has been tested in the present study. One which is commonly applied is based on the uniform flux assumption i.e. $\partial^2 c/\partial n^2 = 0$, where n is the normal to the boundary. While this type of open boundary condition is applicable, the use of it in the present modelling study is not satisfactory. Concentrations at the open boundaries accumulate in time which mask the solution inside the computational domain. While, this could be the result of a highly non-uniform velocity field due to the wind and the tide as used in the model, it could be an indication of partial reflection making it undesirable for use in the present application.

The speed propagation of a variable ϕ (in this case depth-mean sediment concentrations) at the open boundaries can be determined using the general conditions given in Equation (4.34) in the previous chapter. Using these conditions and substituting μ for $c\Delta t/\Delta x$, we obtain for C_L

$$C_L = \frac{c_{B\mp 1}^{n-1} - c_{B\mp 1}^{n+1}}{c_{B\mp 1}^{n+1} + c_{B\mp 1}^{n-1} - 2c_{B\mp 2}^{n}} \qquad (5.51)$$

where the subscript B denotes the boundary. Here, the sediment concentrations at the next two inner grid points at time levels $n+1$, n and $n-1$ are used to get an estimate of the phase speed of propagation of c near the boundary. Using this value C_L in the general condition given by Equation (4.37) gives an estimate of μ. This value of μ is then used to determine the final sediment concentrations at the open boundaries using the general relation;

$$c_B^{n+1} = [c_B^{n-1}(1 - \mu) + 2\mu c_{B\mp 1}^{n}]/(1 + \mu) \qquad (5.52)$$

where the upper sign corresponds to the right open boundary, and the lower sign to the left open boundary. The third order upstream nature of the present numerical scheme implies that two unknown points in the open boundaries have to be specified. This can be easily treated with the present boundary condition by initially solving the value at the second inner grid point then successively solving the value at the open boundary.

This type of open boundary formulation can be used for all open boundaries of the computational domain where sediment concentrations are advected out of the solution domain. In cases where concentrations are discharged into the modelled basin, i.e. from rivers or non-point sources, the sediment discharge from these sources has to be defined from existing measurements or a suitable discharge function must be used. It should be noted that using the proposed open boundary condition for sediment transport modelling produces better results than using the zero-second derivative of concentration (uniform flux) available in literature. Reflection at the boundary is avoided by the present formulation. In several test cases involving unsteady wind and tide-driven flow velocities, concentrations at the boundaries never show any reflection or high fluctuations with the use of the present boundary condition. In such cases, the use of the classical boundary condition (zero-second derivative) produces high and fluctuating concentrations at the boundaries presumably due to reflections of advected concentrations. This can corrupt the solution inside the computational domain especially if the domain is not large such as the Cape Bolinao case wherein boundaries are not far from the main area of interest. The non-uniformity of currents due to the wind and the tide favors the use of the present open boundary condition based on the concept of wave propagation.

Chapter 6

Light Extinction in the Marine Environment

The extinction of light in the marine environment is one of the important water quality variables often addressed by aquatic scientists and oceanographers. The characteristics of the underwater light field itself is a classical subject of oceanographic optics.

In light extinction studies, the Photosynthetically Available Radiation (PAR) corresponding to the wavelength range from 400 - 700 nm is often quantified. Light extinction refers to the diminution of the incident downward irradiance for PAR in the water column. This implies capture of incident PAR by the optically active components present in the water column which may be in the form of dissolved organic substances, living and dead planktonic materials, inanimate suspended solids and water itself.

The availability of downwelling PAR as a water quality variable is mostly related to the survival capacity of underwater living organisms such as submerged flora in the process of photosynthesis, and underwater fauna in their grazing activities. It has a major influence on the growth of phytoplankton and submerged macrophytes (Lijklema et al. 1991, Hootsmans and Vermaat 1991). The available light is not only important for submerged biota but also for human activities in the aquatic medium related to recreation and navigation.

In this study, the importance of light extinction as a water quality variable is addressed in relation to the ecology of the submerged biota in the coastal areas around Cape Bolinao. These light-dependent submerged biota include seagrasses, seaweeds, corals, and many other benthic organisms like clams. Of primary importance is the seagrass population within the vast reef flat around the cape. The available light is one of the primary limiting variables in the growth of this submerged flora, besides nutrients and temperature. Light availability is of major importance not only in determining how much plant growth there will be but also which kind of species will predominate and which kind will evolve (Van Duin 1992). Therefore, a study on the degree of light penetration in the water column and the extinction of the photosynthetically available radiation in such a rich and diversified ecosystem is deemed important.

This chapter presents both theoretical and modelling backgrounds necessary in characterizing the underwater light field. The first three sections give some details on the theoretical background of the subject. Actual measurements available in literature are presented to complement the discussion. The last section deals primarily with modelling light extinction. Several approaches in quantifying light extinction as used in natural waters are presented.

6.1. Introduction

The theory of wave-particle duality of optical physics serves as an important background to deal with the study of the behavior of light which exist as both wave and particle at the same time. This permits both classical geometrical optics and electromagnetic theory to describe the complex behavior of the interaction of light with the particles in a medium (i.e. the aquatic medium in particular). The theories of Rayleigh, Einstein-Smoluchowski, and Mie all deal with the behavior, particularly scattering, of the light particles (referred to as photons or quanta) in a medium. While these theories are successful in predicting the scattering of a quanta of radiation, they can not successfully predict the total diminution of the available light in the aquatic medium due to the complex absorption-scattering mechanism involved. The radiative transfer theory which emerged later became the basis of describing the characteristics of the light field as affected by the inherent optical properties of the aquatic medium. As reported by Kirk (1983), Preisendorfer (1961) has used the equation of radiative transfer to arrive at relations between certain properties of the incident light field and the quasi-inherent optical properties. The inherent optical properties are defined by Preisendorfer (1961) and by Kirk (1983) as the absorption, scattering and beam attenuation coefficients of a medium. The absorption coefficient is defined by Kirk(1983) as the fraction absorbed per unit of path length from a parallel beam of monochromatic light directed normal to an infinitesimally thin layer of medium. Similarly, the scattering coefficient is defined as the fraction scattered of the incident parallel beam divided by the path length. The beam attenuation coefficient is defined as the sum of the absorption and scattering coefficients. An additional inherent optical property describing scattering of photons in a medium is the volume scattering function which defines the angular distribution of the scattered light field.

In the aquatic medium, the real quantities involved are the diffuse absorption and scattering coefficients referred to as quasi-inherent optical properties by Kirk (1983). These quantities are defined analogous to the inherent optical properties but are in general greater in magnitudes because the angular distribution of the incident flux is taken into consideration. This gives greater pathlengths for photons to travel in the medium and hence greater absorption and scattering coefficients. Kirk (1983) gives the equations for the diffuse absorption and scattering coefficients which are functions of depth z as;

$$a(z) = \frac{a}{\overline{\mu}(z)} \quad , \quad b(z) = \frac{b}{\overline{\mu}(z)} \tag{6.1}$$

where $\mu(z)$ refers to the depth-dependent average cosines of all the angles that the quanta of radiation make with the vertical, a is the absorption coefficient, and b is the scattering coefficient.

By definition, the incident light field or downward irradiance in a water column refers to the instantaneous value of the downwelling radiant flux in a horizontal unit area. It is expressed in units of quanta m^{-2} s^{-1}, or more conveniently in μE m^{-2} s^{-1} (1 E = 6.023 x 10^{23} quanta) since a large number of quanta is actually involved. Kirk (1983) differentiates between downward and upward irradiance, the first being that due to downwelling stream of light and the second due to the upwelling stream of light. In light extinction studies, the desirable quantity is the downwelling PAR which is referred to as the downward irradiance covering the 400 - 700 nm range of the wave spectrum. The downwelling PAR is attenuated due to both scattering and absorption processes by the optically active components in the water column. This mechanism of light attenuation by scattering and absorption is further discussed in the following section followed by a discussion in the next section on the factors affecting the actual light extinction process in the marine environment.

6.2. The Attenuation of Downward Irradiance in the Aquatic Medium

The downward irradiance is considered by aquatic scientists to give a good indication of the amount of photosynthetically available radiation in the water column. Its quantification is of primary importance especially when assessing primary productivity in a particular aquatic environment.

The diminution of downward irradiance with depth is approximately exponential in nature such that an incident downward irradiance E_o close the water surface attains a value E_z at depth z in accordance with the Beer's law;

$$E_z = E_o e^{-k_d z} \qquad (6.2)$$

where k_d is the extinction coefficient for downward irradiance. Figure (6.1) shows examples of the diminution of downward irradiance in some natural water systems.

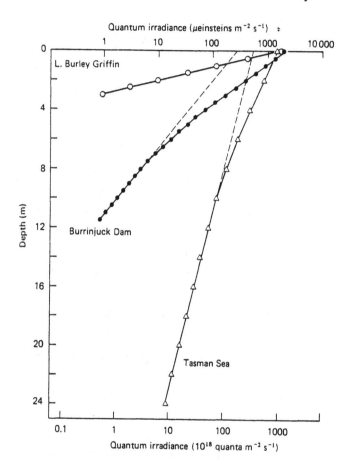

Figure 6.1. The diminution of downward irradiance for PAR with depth (Kirk 1983).

The decrease of the magnitudes of downward irradiance with depth is quite abrupt in both inland waters (Lake Burley Griffin and Burrinjuck Dam). This reflects the strong absorption and scattering of the downward irradiance with depth. In coastal seas, like Tasman Sea, the decrease of irradiance with depth is less due to the relatively weaker absorption and scattering in the marine environment as compared to inland waters. This slow diminution of PAR with depth is related to the lower concentrations of dissolved organic substances and particulate matter in the coastal sea studied.

The diminution of the downward irradiance for PAR is generally observed to follow the exponential relation given in Equation (6.2). However, as the cases of Burrinjuck Dam and Tasman Sea in Figure 6.1 show, the exponential relation is not strictly obeyed in some cases i.e. there is a faster attenuation in the upper few meters of the water column. This is due to the fact that there is a slight decrease of the extinction coefficient with depth in the clearer waters (Tasman Sea and Burrinjuck Dam). The biphasic diminution of PAR irradiance, which is not observed in rather turbid waters as the case in Lake Burley Griffin shows, manifests the dependence of the extinction coefficient with depth. This dependence of the light extinction coefficient with depth, though seldom quantified, is primarily related to the change in the spectral distribution of PAR with depth.

The downward irradiance for PAR varies with wavelength across the photosynthetic range. These differences in the rates of attenuation of downward irradiance as a function of wavelength give rise to a progressive change with depth in its spectral composition. Some field observations given by Kirk (1983) show the spectral distribution of the downward irradiance (see Figure 6.2). In the rather clear waters of Gulf Stream where light penetration is relatively deep, measured quantum irradiance values show that most of the available underwater light is confined within 400 nm to 600 nm. The case at Batemans Bay which manifests a faster diminution of light with depth in coastal areas shows that the irradiance spectra contain high peaks between 550 nm and 650 nm. The quantum irradiance decreases fast towards the blue and the red end of the spectrum which is attributed by Kirk (1983) to attenuation by dissolved substances at the blue end and absorption by water at the red end of the spectrum. These changes in the spectral distribution of downward irradiance within the aquatic medium explain the non-linearity of the logarithm of irradiance when plotted with depth, with consequently a decrease of the extinction coefficient, as shown previously in Figure (6.1).

6.2.1. Absorption and Scattering of Downwelling PAR

Both absorption and scattering coefficients strongly depend on the composition of the aquatic medium. The major light absorbing and scattering components in the water column include dissolved organic substances, dead and living planktonic materials, suspended inanimate particles, and water itself. These components differ in the way they absorb and scatter downward irradiance across the photosynthetic waveband. Measurements of the absorption and scattering coefficients at different wavelengths have shown such variations across the

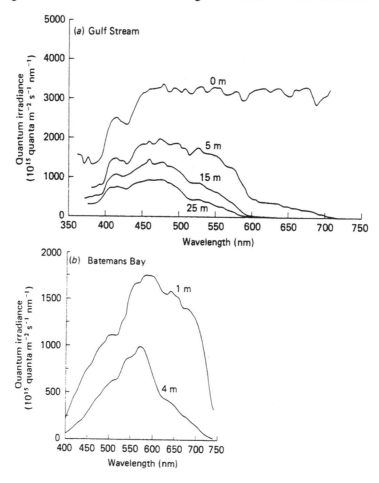

Figure 6.2. Spectral distribution of downward irradiance at different depths in the Gulf Stream and Batemans Bay (Kirk 1983).

PAR range. For absorption alone, typical absorption spectra in various natural waters are shown in Figure 6.3. Note the strong variation of the absorption coefficients within the photosynthetic waveband. Observations in coastal waters show a generally increasing absorption with increasing wavelength (case e). The estuarine observation (case d) shows similar patterns as inland waters, with increasing absorption with decreasing wavelength. These patterns are the result of the total absorption coefficients of all of the major light absorbing components in the water column. The difference in the absorption spectra on both inland and coastal waters are attributed to differences in the nature and composition of the major components present in the water column. Generally, the strong absorption in inland and estuarine waters is attributed to organic substances, gilvin and/or phytoplankton. The different spectral characteristics of the absorption coefficient in coastal waters can be attributed to absorption by water itself. Especially in areas where the concentrations of dissolved or particulate substances are insignificant, absorption by pure water may dominate light extinction.

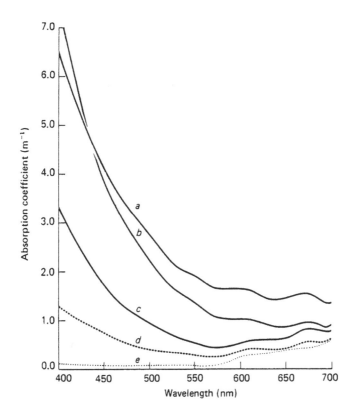

Figure 6.3. Absorption spectra in various natural waters. Cases *a*, *b* and *c* are inland, *d* is estuarine and *e* is marine (Kirk 1983).

The effect of scattering on the other hand is to impede the vertical penetration of light. As pointed out by Kirk (1983), scattering does not itself 'remove' light since a scattered photon is still available for photosynthesis. However, by making the photons follow a zig-zag path, the probability of being absorbed by the absorbing components in the aquatic medium is increased. Hence, with the scattering contribution of suspended particulates for example, the vertical light attenuation is intensified through this mechanism.

6.2.2. The Vertical Attenuation Coefficient

The extinction coefficient k_d in Equation 6.2 is generally used as the vertical attenuation coefficient for downwelling PAR. It is particularly useful to obtain quantitative estimates of the extinction coefficient to determine the extent of primary production due to light limitation in surface water systems. As pointed out by Kirk (1983), it is the best single parameter by means of which different water bodies may be characterized in terms of the availability of the photosynthetically useful radiant energy within them. It is also useful when considering general water quality characterization of natural water systems.

The vertical attenuation coefficient for downwelling PAR is generally a measure of the energy 'remove' from an incident light field as it penetrates the water column. This 'removal' of energy is generally a non-linear function of the absorption and scattering mechanism attributed to the character of the light field itself and the inherent and quasi-inherent optical properties of the aquatic medium. While the extinction coefficient can be estimated from the general relation given by Equation (6.2), its precise quantification should be determined by the general relation obtained by Preisendorfer using radiative transfer theory;

$$k_d(z) = a_d(z) + b_{bd}(z) - b_{bu}(z)R(z) \qquad (6.3)$$

where $k_d(z)$ is the vertical attenuation coefficient for downward irradiance taken as a function of depth, $a_d(z)$ is the diffuse absorption coefficient for downward irradiance, $b_{bd}(z)$ the diffuse backscattering coefficient for downward irradiance, $b_{bu}(z)$ is the diffuse backscattering coefficient for upward irradiance and $R(z)$ is the irradiance reflectance which is the ratio of the upward to the downward irradiance ($= E_u/E_d$). Note that all the terms involved depend on the depth z. In general, the use of Equation (6.3) is difficult to apply since direct measurement of the involved variables i.e. the diffuse scattering coefficients, is not yet feasible. The use of the normal absorption and scattering coefficients was found possible instead by Kirk (1981). The vertical attenuation coefficient, assumed constant with depth, is found to fit the general non-linear relation given by

$$k_d(z_m) = (a^2 + 0.256ab)^{1/2} \qquad (6.4)$$

where z_m denotes the mid-point of the euphotic zone, the layer at which PAR falls to 1 % of that just below the surface. In this layer, most of the photosynthesis takes place. Usually, measurements of the absorption and scattering coefficients are difficult to execute, hence the use of Equation (6.4) is also not a practical way of determining the vertical attenuation coefficient for downwelling PAR. The availability of commercial equipment for measuring the downward irradiance throughout the photosynthetic waveband allowed the direct application of Equation (6.2) in the determination of k_d. Taking logarithms of both sides and rearranging gives;

$$k_d = \frac{1}{z} \ln\left(\frac{E_o}{E_z}\right) \tag{6.5}$$

In principle, only two values at different depths of the downward irradiance are needed to estimate the extinction coefficient, namely E_o taken at any depth in the water column and E_z taken at any depth z below E_o. However, k_d is a function of the water depth (see Figure 6.1). Observations showed that in some inland and coastal waters, the extinction coefficient slightly decreases with depth. As absorption of PAR decreases with depth, k_d is also expected to decrease with depth. On the other hand there is an amplification of the photon pathlengths with depth because light becomes increasingly more diffuse with depth (Kirk 1983). In most observations, the extinction coefficient is taken to be approximately constant with depth. The near-constancy of the extinction coefficient with depth can be due to two opposing tendencies. The progressive removal of the more strongly absorbed wavelengths gives decreasing k_d values with depth, but the angular distribution becomes more diffuse with depth giving increasing k_d values at the same time. It may be that in the more turbid waters which show constant k_d values with depth, these two tendencies cancel each other out (Kirk 1977).

6.3. Factors Affecting Light Extinction in the Marine Environment

Light extinction in natural waters is affected by four primary groups of substances whose composition and concentration differ in each water body giving different values of the extinction coefficient in each of these water bodies. Furthermore, the extinction coefficient may change with time due to the varying composition and concentrations of the primary factors. These factors, which are referred to as optically active components of the water column, include inanimate suspended solids, dead or living phytoplankton (algae), gilvin, and water itself. The extent by which each of these factors affect the light penetration differ from each other due to the fact that the absorption and scattering characteristics of these components are different. The light attenuating characteristics of these components generally vary with wavelength λ across the photosynthetic range. Table (VI.1) gives a short description of the absorption and scattering characteristics of the different components.

Component	Absorption	Scattering
Tripton	very weak close to the red end; decreases with increasing λ	strongest relative to other components
Algae	peaks at 440 and 670 nm; function of cell size and shape	less significant compared to absorption
Gilvin	strong at low λ; decreasing exponentially with increasing λ	not significant; no reported value
Water	very weak in the blue/green region; significant above 550 nm;	very low value of 0.0058 at 400 nm

Table VI.1.   Light attenuating characteristics of the different optically active components.

The observed absorption and scattering efficiencies (measured in terms of a and b) of these components may vary for different water bodies due to differences in particle size distribution or composition (Blom et al. 1994). A more detailed description of the attenuation characteristics of these components is provided in the following sections.

6.3.1. Inanimate Suspended Solids

The inanimate suspended solids referred to as tripton contributes strongly to the extinction of downwelling PAR through the combined absorption and scattering mechanisms (see Table VI.1). Especially in shallow coastal waters adjacent to estuaries where concentrations of suspended particles are substantial, the vertical light penetration can be reduced significantly. At typical concentrations however, this particulate fraction does not absorb light strongly but scatters quite intensely (Kirk 1983). Due to the very low settling velocities of some of these particles, they remain suspended in the water column and for a long time contribute to the overall diminution of PAR irradiance. The effect of increased concentrations on the extinction coefficient is, as shown by observations, approximately linear, i.e. $k_{d,c} = \kappa c$, in which $k_{d,c}$ is the extinction coefficient due to the suspended matter, κ is a specific extinction coefficient and c is the concentration.

Some examples of the increase of absorption with decreasing wavelengths for inland waters are presented in Figure 6.4. The absorption spectra sometimes show two peaks (one at about 440 nm and another at 670 nm). These observed peaks are related to the presence of suspended algae which are difficult to separate physically from the inanimate suspended particles.

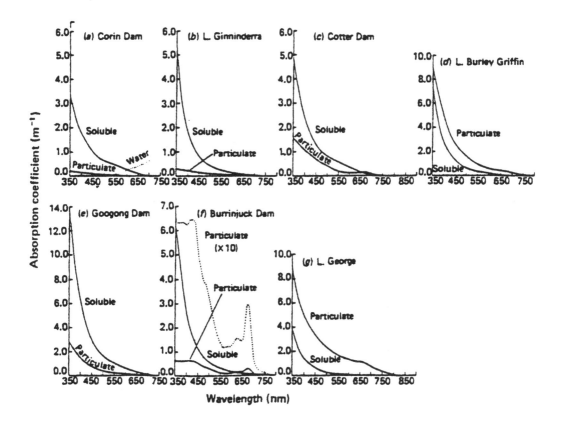

Figure 6.4. Absorption spectra of suspended particles in some inland waters (Kirk 1983).

As said, scattering generally does not 'remove' light but increases the pathlengths of the photons and therefore the probability that these photons are absorbed. The Mie theory of scattering, which is applicable to particle sizes even larger than the wavelength of light, predicts that most of the scattering is in the forward direction. Available measurements on the turbidity of natural waters expressed in nephelometric turbidity units (NTU) give some indication on the scattering efficiency of suspended particles in certain natural waters. The measured turbidity in this way, where 1 NTU roughly corresponds to a scattering coefficient $b = 1$ m^{-1} (Kirk 1983), is due to both inanimate suspended solids and phytoplankton. However, in unproductive waters where phytoplankton concentration is negligible, it is mainly due to the inanimate fraction.

The size of suspended particles is an important factor in the light extinction. Baker and Lavelle (1984), in an attempt to determine the effect of particle size (mass concentration) on the light attenuation coefficient of natural suspensions, found that suspensions with a mean particle size of 8.5 μm attenuate 660 nm light 15 times more efficiently than suspensions of similar particles with mean diameter of 48 μm. Observations in both freshwater and marine waters also indicate that the specific extinction coefficients of suspended particles increase with decreasing size fraction (Kirk 1983, Van Duin 1992). This further implies that the contribution of the smaller particles to light attenuation is greater (per unit weight). Theoretically, the contribution to light extinction by these inanimate suspended particles is proportional to their cross-sectional area, and not to their volume (Van Duin 1992, Blom et al. 1994). With similar concentrations, (expressed as mass per unit volume) this implies higher extinction coefficients for clay particles than silt and sand, a common observation in natural waters.

6.3.2. Phytoplankton

Partly, the penetration of light into the water column is affected by the presence of dead and living phytoplankton. These planktonic materials both absorb and scatter PAR resulting in reduced light penetration. In particular, the absorption of PAR by the algal cells can impart a substantial decrease of the available light in the water column (see Table VI.1). The photosynthetic pigments of the phytoplankton community which include chlorophylls, carotenoids and biliproteins, all contribute to absorption of PAR (Kirk 1983). This is especially important in productive inland and coastal waters, and also in areas of upwelling in the coastal ocean. The degree of PAR absorption by phytoplankton depends not only on their concentration but also on the size and shape of the algal cells (Kirk 1983). In his study, Kirk (1977) found that the absorption coefficient due to phytoplankton can be approximated by $a_c = N \times A$, where a_c is the absorption coefficient due to phytoplankton, N is the number of algal cells per unit volume, and A is the absorption cross-section which varies (throughout the photosynthetic waveband) for different algal species due to differences in shapes and sizes. Typical spectral variations of the mean absorption cross-section of several sizes and shapes of blue-green algae are presented in Figure 6.5. The absorption cross section, which tells something about the amount of light of a given wavelength that a single cell or colony takes out from a beam of unit area, clearly depends on the sizes and shapes of algal cells present. According to Kirk (1977), the mean absorption cross-section differs for diatoms, dinoflagellates, green, or blue-green algae. The blue-greens have been recognized to contribute more to the attenuation of PAR than any other algae.

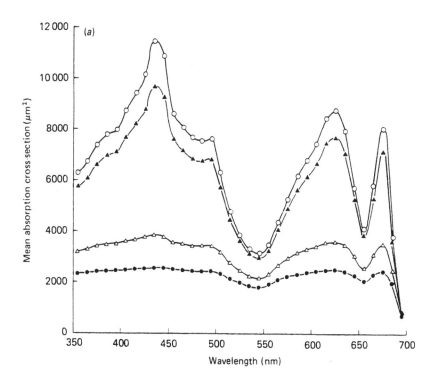

Figure 6.5. Spectral variation of the mean absorption cross-section of various sizes and shapes of blue-greens. The topmost curve refers to 6 μ diameter spheres; second curve - 6 μ cylinders; third curve - 28.8 μ prolate spheroids; lowest curve - 57.6 μ diameter spheres (Kirk 1983).

The specific absorption coefficient of marine phytoplankton (absorption coefficient per unit concentration) is typically represented by two absorption peaks within the photosynthetic range. The spectral variation of the specific absorption coefficient obtained by Morel and Prieur (1977) from measured light extinction and scattering coefficients is documented by Kirk (1983). Figure 6.6 shows the estimated spectral variation of the specific absorption coefficient for marine phytoplankton obtained at the Atlantic off Northwest Africa. The spectral variation, corresponding to 1 mg chlorophyll a per m^3 for oceanic phytoplankton, shows a higher peak at 440 nm and a lower one between 600 and 700 nm. These correspond roughly to the spectral variation of the mean absorption cross-section of the phytoplankton studied (see Figure 6.5).

The total contribution of a certain phytoplankton community to light extinction expressed by the partial extinction coefficient due to algae can be determined from the specific extinction coefficient and the total number of algal cells. Van Duin (1992) found a general linear

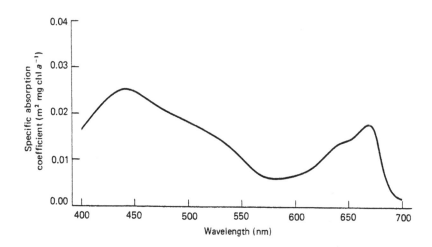

Figure 6.6.   Typical specific absorption spectra for marine phytoplankton (Kirk 1983).

relationship in the form: $k_{d,a} = \kappa_a \cdot N$, where $k_{d,a}$ is the light extinction coefficient due to algae, and κ_a is in this case a specific extinction coefficient (extinction coefficient per unit algal cell of *Oscillatoria agardhii*). The estimated total extinction coefficient which included the algal extinction estimated in this way was very close the observed extinction coefficient (Van Duin 1992).

6.3.3. Gilvin

One of the major optically active components in the water column is gilvin (gelbstoff). Gilvin is dissolved organic matter, composed chiefly of humic and fulvic acids (Carder et al. 1989). It is generally recognized that these humic and fulvic substances (which differ in molecular weights) originate from decomposed plant tissues in upland soil or in natural waters (Kirk 1983). In the marine environment, the highest concentrations of gilvin occur in regions influenced by land drainage, such as the Baltic Sea and several coastal waters (Carder et al. 1989). Furthermore, as reported by Kirk (1983), Kopelevich and Burenkov (1977) observed that gilvin concentration is strongly related to the level of phytoplankton chlorophyll in productive oceanic waters.

Gilvin strongly absorbs incident light at low wavelengths in surface water systems and is generally responsible for the yellow colors in the water column, particularly when its

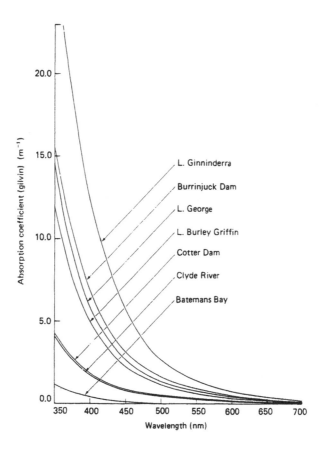

Figure 6.7. Typical absorption spectra of gilvin containing Australian natural waters. The lowest curve shows a typical observation for a marine environment (Kirk 1983).

concentration is high enough such as in certain inland waters. The profound effect of gilvin on the absorption of light especially near the blue end of the PAR spectrum is well documented in both inland and coastal waters (see Table VI.1). Absorption of downwelling PAR by gilvin decreases with increasing wavelength. Figure 6.7 gives some typical spectral variations of the absorption coefficients in both inland and marine waters.

High concentrations of gilvin in inland waters give very high absorption coefficients especially below 550 nm as observed from the figure. In marine waters such as the one considered here (Batemans Bay), very low concentrations of gilvin impart low absorption. The general absorption pattern throughout the photosynthetic waveband shows a general exponential decrease with increasing wavelength. With respect to the peak-absorption wavelength λ_0, the absorption coefficient at any wavelength λ can be estimated from the approximate relationship proposed by Bricaud et al. (1981) as presented by Kirk (1983)

$$a(\lambda) = a(\lambda_o)e^{-0.014(\lambda - \lambda_o)} \qquad (6.6)$$

where $a(\lambda)$ and $a(\lambda_o)$ are the absorption coefficients at wavelengths λ and λ_o (in nm) respectively. Normally, λ_o is taken at lower wavelengths (440 nm or 380 nm) where absorption is strongest and hence easily measured.

The absorption of light by gilvin throughout the photosynthetic waveband has significant influence on the ecology of aquatic systems. It amounts to direct competition with phytoplankton and other aquatic plants for capture of available light energy (Kirk 1977; Kirk 1983). Also, because of greater absorption of short than long wavelength light, it affects vision of aquatic fauna by imparting yellow color to the water (Davies-Colley and Vant 1987). While this is especially true in inland waters such as lakes and rivers, it is also possible in coastal waters influenced by land drainage and relatively rich in marine flora giving high concentrations of gilvin. The resulting light extinction can be dominated by gilvin absorption in these surface water systems.

6.3.4. Water

The incident light in the marine environment is attenuated as well by 'pure' sea water. In the absence of particulate or dissolved substances in the marine environment, light attenuation still exists due to the influence of the water molecules on the vertical penetration of photons. This attenuation is the result of both absorption and scattering processes although absorption is the more prominent mechanism (Table VI.1).

In contrast to gilvin, the absorption by pure sea water is stronger at higher wavelengths. Absorption is very weak and very difficult to measure in the blue-green spectral region (below 500 nm). However, absorption begins to rise as wavelength increases above 550 nm and is quite significant in the red region (Kirk 1983). It was estimated by Kirk (1983) that a 1 m thick layer of pure water will absorb about 35 % of incident light of wavelength 680 nm. The absorption spectrum of pure sea water has been estimated by Smith and Baker (1981) based partly on measurements of the extinction coefficient in clear oceanic water and partly on laboratory measurements. Figure 6.8 shows the spectral variation of the absorption coefficient of pure sea water within the photosynthetic waveband. Averaging the observed values over the wavelength from 400-700 nm gives a mean absorption coefficient of 0.165 m^{-1}, with a low value of 0.0171 m^{-1} at 400 nm increasing non-linearly to 0.650 m^{-1} at 700 nm.

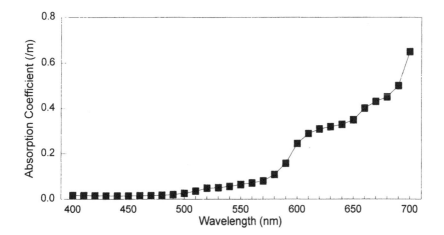

Figure 6.8. Spectral variation of the absorption coefficient of pure sea water. Data were taken in a clear ocean by Smith and Baker (1981) and compiled by Kirk (1983).

Scattering of light by water is observed to be very weak. As reported by Kirk (1983), Morel (1974) found very low values of the scattering coefficient for pure water (0.0058 m^{-1} at 400 nm) indicating that scattering by pure water does not contribute much to light attenuation especially within the photosynthetic range.

6.4. Modelling Underwater Light Extinction

The underwater light field is diminished by the combined absorption and scattering efficiencies of all the optically active components in the water column. The resultant extinction coefficient reported in literature is influenced not only by the quality of the incident downward irradiance for PAR, but the concentrations in the water column of the factors just considered. Furthermore, the composition of both dissolved and suspended materials in the water column also affects the overall diminution of PAR.

Modelling the underwater light field requires some information on the preceding factors. The usual procedure is to quantify the extinction coefficient k_d which could vary in space and time in a particular water body. In an attempt to obtain a quantitative description of the light field in natural waters, Kirk (1984) used a Monte Carlo approach by considering the spectral variation of the absorption and scattering coefficients in these waters. The general extinction relation given by Equation (6.4) is modified to take into account the effect of the angle of incident light field (zenith angle) in the water column represented by the average cosine μ. This modified relation for the extinction coefficient, which includes variations within the photosynthetic waveband is given by;

$$k_d(\lambda) = \frac{1}{\mu}[a(\lambda)^2 + (0.425\mu - 0.19)a(\lambda)b(\lambda)]^{1/2} \qquad (6.7)$$

where λ is the wavelength, μ is the average cosine of the zenith angle, and a and b are the (diffuse) absorption and scattering coefficients respectively as defined in Equation (6.1). From this relation, the average extinction coefficient of PAR can be obtained by integration of the resulting spectral variation of k_d within the photosynthetic range. While Equation (6.7) may give realistic result for the light extinction coefficient in natural waters, the problem is transposed to the determination of both scattering and absorption coefficients which is not feasible. Therefore, such a relation is not often used in modelling the underwater light extinction.

A common method often employed in modelling the extinction of downward irradiance is to consider the influence of the major optically active components separately giving partial extinction coefficients for each component. The sum of all the partial extinction coefficients gives the average extinction coefficient of the water column in question. The additive approach is represented by;

$$k_d = k_{d,w} + k_{d,p} + k_{d,c} + k_{d,g} \qquad (6.8)$$

where k_d is the total extinction coefficient, $k_{d,w}$, $k_{d,p}$, $k_{d,c}$ and $k_{d,g}$ are the partial extinction coefficients of pure water, suspended particulate matter, phytoplankton chlorophyll and gilvin respectively. The general method consists in determining the partial extinction coefficients of the major factors considered. The partial coefficients generally can be determined from the specific extinction coefficient and the concentration of materials in the water column by the relation;

$$k_{d,n} = \kappa_n c_n \qquad (6.9)$$

where $k_{d,n}$ is the extinction coefficient of a particular component, κ_n the specific extinction of that component, and c_n the observed concentration. Using this, the general equation for the total extinction coefficient can be written as;

$$k_d = k_{d,w} + \kappa_p c + \kappa_c Chl\text{-}a + \kappa_g G \qquad (6.10)$$

where κ_p, κ_c, and κ_g are the specific extinction coefficients of suspended particulate, phytoplankton and gilvin respectively, c is the concentration of suspended particulate, *Chl-a* is the phytoplankton concentration taken as chlorophyll-a, and G is the absorption coefficient of gilvin usually measured at lower wavelengths (i.e. 380 nm) where significant absorption exists. Field measurements and/or model predictions of the concentrations of suspended particles, phytoplankton measured as chlorophyll-a, and gilvin absorption are essential to get a quantitative estimate of k_d. Regression analyses is often used to estimate the specific extinction coefficients of each component. The partial extinction coefficient of water is used as a background coefficient which may be assumed constant (literature value).

The use of Equation (6.10) can be further decomposed for the particulate fraction to take into consideration the effect of particle size classes. With the three sediment fractions of interest (sand, silt and clay), this gives;

$$k_d = k_{d,w} + \kappa_1 c_1 + \kappa_2 c_2 + \kappa_3 c_3 + \kappa_c Chl\text{-}a + \kappa_g G \qquad (6.11)$$

where κ_1, κ_2, κ_3 are the specific extinction coefficients of sediment fractions 1 to 3, and c_1, c_2, c_3 are their respective concentrations.

Generally, the suspended particulate in the water column are composed of varying size fractions imparting different degrees of light extinction. The use of an extended light extinction model taking into account the contribution of different size fractions of suspended particulate and all the major components has been successfully applied in inland waters systems. The application of such a relation in the coastal sea is generally valid since the major optically active components are similar, although concentrations may be lower. The general relationship given by Equation (6.11) is therefore used in this study for modelling light extinction around Cape Bolinao. The estimation of the specific extinction coefficients as used in the light extinction model is described in the following chapter.

Chapter 7

Field Observations and Laboratory Experiments

The field observational study and laboratory measurements were undertaken for a period of nearly two years. Although some measurements started in August 1993 due to the earlier availability of certain equipment, the full operational observational study started at the beginning of the year 1994. All measurements in both field and laboratory ended in June 1995. It was necessary to execute field research and laboratory measurements for longer periods in order to obtain information on the seasonal variability of the physical processes involved.

The analyses of the measured variables, using available computer programs was completed only at the beginning of 1996. Some analyses, like tidal analyses had to be done with newly developed computer programmes due to unavailability of special packages for it. However, most of the data analyses were carried out using available statistical programs. Since not much research studies related to the processes investigated here were available for the study site itself, comparison with previous data is limited. This study thus is particularly important for understanding the relevant physical processes of the marine environment around Cape Bolinao. With the aid of the modelling studies already described in the previous chapters, the marine physics of the coastal zone within the area of study is extensively characterized in this research. The following sections describe the results of the field and laboratory work, with the use of appropriate tools for analyses where necessary, in the marine waters off Cape Bolinao. It should be noted that the data from uninterrupted simultaneous measurements of relevant variables are particularly useful in the analyses.

7.1. Hydrodynamics off Cape Bolinao

The current patterns within the reef flat and surrounding areas of Cape Bolinao where most of the measurements were executed are very complex due to the interactions of several significant factors including wind and tidal forces and general topography. The wind may dominate the flow pattern in some instances due to the shallowness of the reef waters. The wind stress generates drift currents due to the long wave circulation phenomena as discussed in Chapter 4. In addition to this, there is the astronomical tide whose propagation from the open sea to the coastal zone gives rise to tidal currents. The periodic ebbing and flooding can influence to a large extent the general circulation pattern within the reef flat. It is essentially the interaction of the wind and tide-generated currents that govern the general circulation pattern off Cape Bolinao.

The effect of the wind is not only confined to the development of the long-wave currents. Wind-induced surface waves may affect the observed current patterns within the reef flat. The radiation stresses due to the spatial variations in the momentum contained by these waves may introduce wave-generated circulation. This phenomenon, including the combined processes of wave breaking, shoaling, refraction and diffraction, may modify to some extent the long wave-induced circulation pattern. Additionally but less significant, the effects of atmospheric pressure gradients and density differences induced by freshwater input and net radiation, may be seen as external forces that can have some influence on the general circulation pattern.

As evidenced by current measurements, however, the general circulation off Cape Bolinao is primarily governed by the wind and tide-driven long-wave currents. The temporal change in the wind speed and direction, coupled with the spring-neap cycle of the tide within the area dictate to a large extent the observed circulation pattern. The seasonal variation in the wind pattern, and the mixed tidal variations give rise to a dynamic circulation pattern whose strength and persistence are partly dictated by the complex topography and bathymetry of the area.

7.1.1. Currents

Measured surface currents during the whole period of observation have been averaged for the southwest and northeast monsoon respectively (see Figure 7.1). Generally, these averaged surface currents were less than 10 cm s^{-1} at most sites. This low value is typical for the shallow waters of the reef during ordinary conditions (no storms) as most measurements of surface currents are done during such conditions. The surface current patterns for the southwest monsoon season reveal a general eastward mass transport pattern. The case for northeast monsoon shows a general southward transport. Both observations illustrate the significant influence of the wind stress on the general current pattern in the shallow reef. Southwesterly to westerly winds, which prevail during the southwest monsoon, drive currents in the eastward direction. On the other hand, northwesterly to northerly winds which prevail during the northeast monsoon, drive currents in the general southward direction. As shown by meteorological observations, the prevailing wind during the northeast monsoon is northwesterly to northerly, and not northeasterly as expected. This is due to the blocking of northeast winds by the mountain ranges (of western Luzon) northeast of the study site.

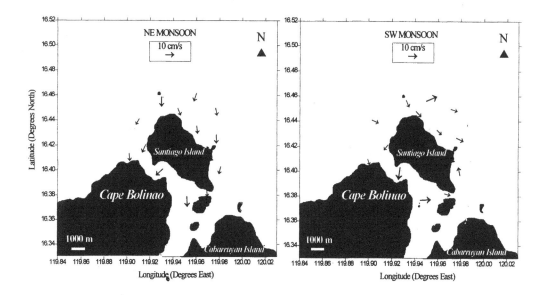

Figure 7.1. Average surface currents at the Bolinao reef for northeast and southwest monsoon seasons.

The effect of the tide is less obvious in the observed surface current patterns, as surface currents are primarily wind-driven. However, at some sites like the channel between mainland Bolinao and Santiago Island, the flows are largely dictated by tidal flooding and ebbing. The effect of the tide can be seen in Figure 7.2 which shows a typical current measurements by a self-recording current meter deployed at the channel between Santiago Island and the mainland. The observed time series of current is decomposed into east-west and north-south components (*u* and *v*). A distorted tidal ellipse is observed, rotating clockwise with the major axis approximately oriented along the channel direction. The irregularity in the tidal ellipse is due to the coastal geometry and bottom topography.

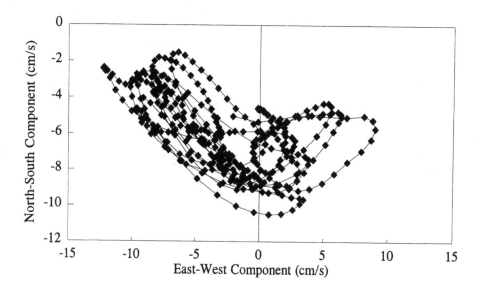

Figure 7.2.   East-west and north-south current components of hourly-observations at the channel bordering Cape Bolinao and Santiago Island during the period 27 Jan - 12 Feb, 1994.

This observation shows that the tidal current penetrates eastward inside the channel during flood tide. At ebb tide, the current is in the westward direction. Generally, the flood current (< 10 cm s^{-1}) is weaker than the ebb current (≥ 10 cm s^{-1}) due to tidal attenuation. Frictional influence due to decreasing water depth from west to east of the channel, combined with coastal obstruction on the tidal propagation is generally responsible for this observation.

Analysis of the observed current and wind (measured in intervals of several minutes and averaged hourly) shows the significant influence of the wind in driving the currents at Cape Bolinao. An hourly-observation of wind and current during the period 8-20 December 1993 is shown in Figure 7.3. Prior to analysis, smoothing of the data is applied. The resistant non-linear smoothing option (5-point) in the statistical package STATGRAPHICS is applied (also throughout the analysis presented here) since this gives a (smoothed) time series that is more resistant to outliers than simple moving averages. A cross-correlation analysis of the smoothed data reveals positive cross-correlation coefficients (Figure 7.4). The estimated coefficients confirmed the significant influence of the wind on the observed current.

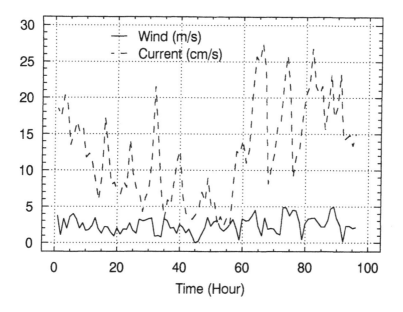

Figure 7.3. Time series of wind and current at the Bolinao reef during the period 8-20 December 1993.

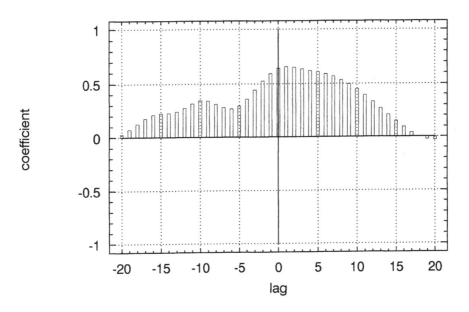

Figure 7.4. Cross-correlation coefficients of wind and current during the period 8-20 December 1993. The time (differential) lag is a multiple of the sampling interval (1 h).

During the southwest monsoon season of 1994, the winds of a tropical storm (codename

'Iliang') with its center located west of the study area, had a profound effect on the currents off Cape Bolinao. Strong southerly winds with surface winds approaching 20 m s^{-1} were experienced in the Bolinao reef on 10 July 1994 (Figure 7.5). Results of the cross-correlation and spectral analyses of the wind and current data for this event are shown in Figure (7.6a,b). Both analyses showed the strong influence of the wind on the currents during this period.

During ordinary conditions (calm to moderate winds), analysis of wind, current, and tide data showed the relative influence of the wind and the tide on the observed current in the study area. A simultaneous observation of wind, current and tide near the Bolinao coast during the period from 23-31 May 1994 is shown in Figure 7.7. A cross-correlation analysis of the time series were performed to quantify the influence of the wind and the tide in driving the currents off Cape Bolinao (Figure 7.8). With winds of less than 10 m s^{-1} during this period, the tidal influence is shown to be more significant (Figure 7.8b).

A simultaneous observation of the wind, current and tide in the Bolinao reef in another ordinary condition (no storm) during the period 23-29 June 1994 is also shown in Figure 7.9. A cross-correlation analysis of the time series suggests the significant influence of the tide in the observed current (Figure 7.10b). The wind is shown to be of minor importance during this period (Figure 10a).

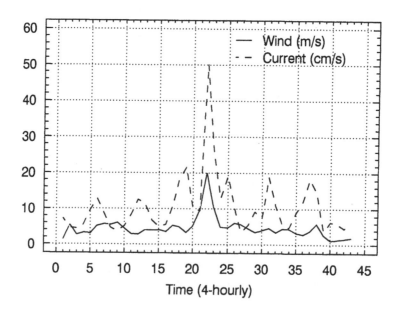

Figure 7.5. Time series of wind and current in the Bolinao reef during a stormy period in July 1994.

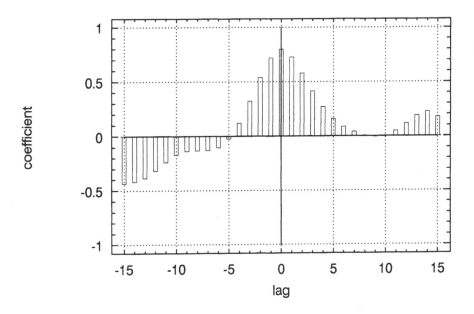

Figure 7.6a. Cross-correlation analysis of the time series in Figure 7.5. The time (differential) lag is a multiple of the sampling interval (4 hours).

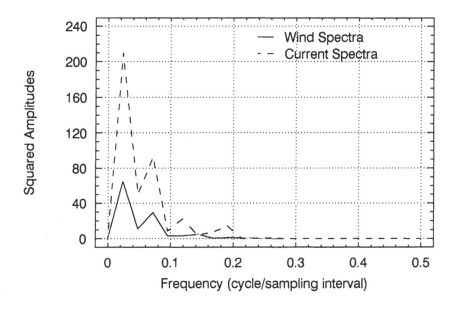

Figure 7.6b. Spectral analysis of the time series in Figure 7.5.

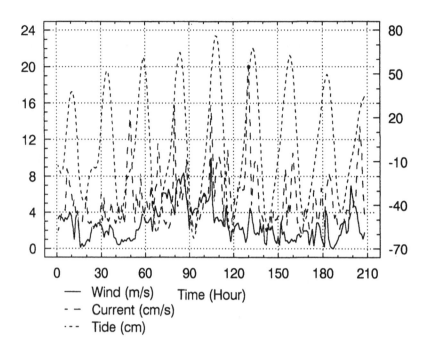

Figure 7.7. Time series of wind, tide and current in the study area during the period 23-31 May 1994.

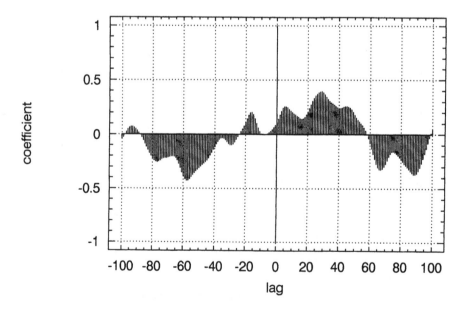

Figure 7.8a. Cross-correlation coefficients for the time series of wind and current in Figure 7.7. The time (differential) lag is a multiple of the sampling interval (1 hour).

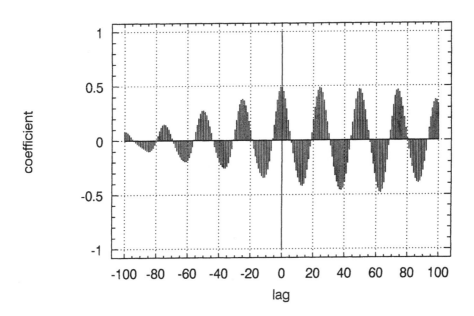

Figure 7.8b. Cross-correlation coefficients for the time series of tide and current in Figure 7.7. The time (differential) lag is a multiple of the sampling interval (1 hour).

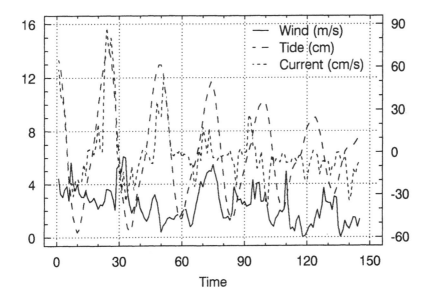

Figure 7.9. Time series of wind, tide and current in the study area during the period 23-29 June 1994.

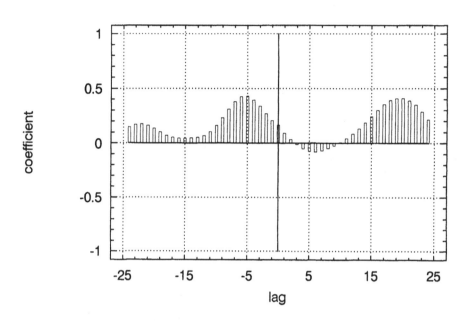

Figure 7.10a. Cross-correlation coefficients for the time series of wind and current in Figure 7.9. The time (differential) lag is a multiple of the sampling interval (1 hour).

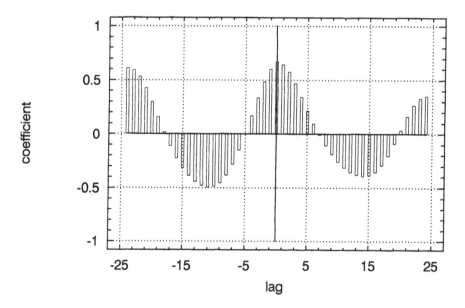

Figure 7.10b. Cross-correlation coefficients for the time series of tide and current in Figure 7.9. The time (differential) lag is a multiple of the sampling interval (1 hour).

Evidently, during ordinary (stormless) conditions, the influence of the tide becomes more dominant than the influence of the wind in driving the current in the study area. Generally however, there is a non-linear interaction between the wind and the tide-driven current. Depending on the tidal regime and the magnitude of the wind speed (and direction) in the area, either of the two driving forces may be more significant. Due to the shallowness of the reef waters, the wind stress is expected to dominate the flow during enhanced wind speeds especially during neap tides when tidal currents are weak. The tidal effect on the other hand, can be very significant especially during spring tide with tidal ranges exceeding 1 m. With calm to moderate conditions in such occasions, it is shown that the tide will dominate the currents of the study area.

7.1.2. Tides and Water Level Variation

The observed tide at the study site is mixed diurnal-semidiurnal with a dominant diurnal characteristic. The O_1 (diurnal principal lunar) and K_1 (diurnal luni-solar) tides with periods of 25.8 h and 23.9 h respectively, are generally the main constituents responsible for the diurnal characteristic of the tide in the area. On the other hand, the M_2 (semidiurnal principal lunar) and S_2 (semidiurnal principal solar) tides with periods of 12.4 h and 12.0 h respectively, are the main semi-diurnal constituents involved. Observations indicate that the diurnal constituents dominate over the semi-diurnal constituents. This was also confirmed by a harmonic analysis of tidal observations which consistently gave higher amplitudes for the diurnal constituents. Table VII.1 shows the amplitudes and phases (including the frequencies) of the four principal tidal constituents derived by harmonic (Fourier) analysis of the observed tide in May-June 1994 (see Figure 4.4) using the known frequencies ($\omega = 2\pi$/period) of the constituents involved. The derived amplitudes of the O_1 and K_1 tidal constituents are approximately 0.21 and 0.33 m respectively, while the corresponding amplitudes of the M_2 and S_2 tides are approximately 0.12 and 0.04 m respectively.

Tidal Constituent	Frequency (rad/hr)	Amplitudes (m)	Phases (radian)
O_1	0.243351	0.2136	2.053720
K_1	0.262516	0.3372	-0.275370
M_2	0.505868	0.1232	3.142611
S_2	0.523598	0.0386	-0.897434

Table VII.1. Frequencies (rad h⁻¹), amplitudes (m) and phases (rad) of the four principal tidal constituents (O_1, K_1, M_2 and S_2) in the area of study.

7.1.3. Temperature and Salinity

Time series of recorded temperature (measured in intervals of 5 minutes) at the reef shows a diurnal fluctuation with an average range of about 2 °C. An example of this observed fluctuation can be seen in Figure 7.11.

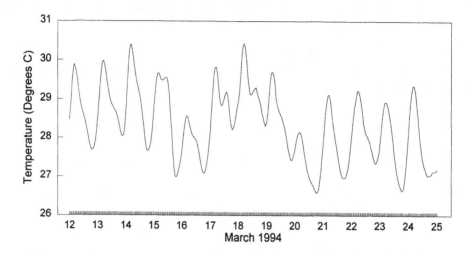

Figure 7.11. Diurnal fluctuation of temperature in the reef waters off Cape Bolinao.

The general daily variation in solar heating (insolation) coupled with the shallowness of the reef waters give rise to this general diurnal fluctuation of temperature. Measurements (at mid-depth) in deeper areas such as the channel separating Santiago Island and Cape Bolinao ($h > 20$ m) does not show such a substantial range in the diurnal fluctuation of temperature.

On a longer time scale, a seasonal variation in temperature can be observed. Weekly observations in the reef and in the channel for more than a year show typical seasonal variations in the temperature (Figure 7.12). The low water temperature in December starts to rise in February and attains its peak in April (summer). A decreasing temperature is observed during the start of the southwest monsoon season. The onset of the rainy season is responsible for this cooling. However, another peak is observed within this season particularly between August and September. This second peak, though lower than the summer peak is similarly attributed to an increase in surface heating during this period. The irregularity in this peak is attributed to the passage of some low pressure areas. It should be noted that these atmospheric disturbances are experienced during the southwest monsoon (see Chapter 3). Broken clouds, which decrease insolation, coupled with intermittent rains during such weather conditions contribute to temperature fluctuations within the shallow reef waters.

Observations of salinity during a period of less than a year (August 1994 - May 1995), indicate low salinity at the reef (S < 32) during the southwest monsoon season (Figure 7.13). Increased precipitation and low evaporation during this season are factors contributing to such low salinity values in the measurement sites. Observations by McManus et al. (1992)

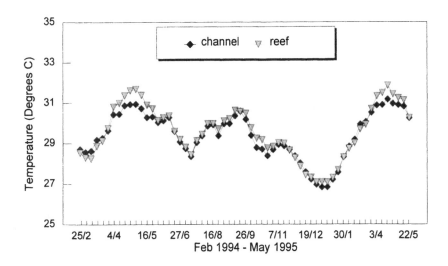

Figure 7.12. Seasonal variation of water temperature in Cape Bolinao.

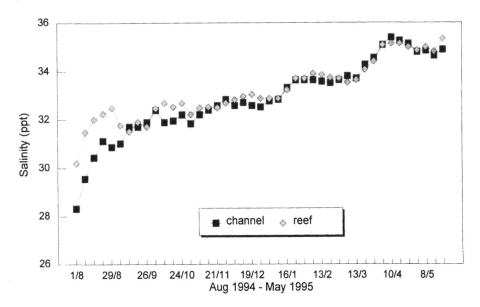

Figure 7.13. Temporal variation of salinity in Cape Bolinao during the period from August 1994 to June 1995.

showed even lower salinity values (S < 30) in the reef waters during strong precipitation. Increasing salinity values are observed towards the northeast monsoon season. This attains its peak in the summer particularly in April (S > 34), the period of almost zero precipitation in the area. The very low input of freshwater during the summer plus enhanced evaporation leads to increased salinity values during this period.

7.2.   Sediment Transport in Cape Bolinao

Sediment transport in Cape Bolinao can be quite complex due to the complexity of the associated hydrodynamics responsible for transport.   There is a selective transport of sediments in a given hydrodynamic regime owing to differences in the erodability of bottom particles and settling characteristics of the suspended particles. Fine sediments require lower bottom shear stress for bed erosion and resuspension in the water column than coarse sediments. At the same time, fine sediments have low settling velocities. This means that silt for example, can be transported already during moderate conditions (ordinary breeze). On the other hand, coarse sediments like sand remain on the bottom during such ordinary conditions with low bottom stress, and can only be eroded and transported during extreme conditions, e.g. during the passage of storms.

7.2.1. Dynamics of Suspended Sediments

Analyses of total suspended solids (TSS) concentrations, current and wind during a tidal cycle and a period of several weeks to months revealed the relative influence of both current and waves in the dynamics of suspended sediments in Cape Bolinao. A cross-correlation analysis of a time series obtained during 9-20 December 1993 and 16-26 June 1994 showed that during ordinary conditions (no storms), the current has a more significant influence on the total suspended solids concentration (Figure 7.14a). The influence of the wind is shown to be insignificant (Figure 7.14b). The negative cross- correlation coefficients obtained between wind and TSS indicate an insignificant influence from the wind during ordinary conditions. This is also evident in the data obtained during the period 12-25 March 1994 in the area north of the Bolinao reef (Figures 7.15a - 7.15b).

A spectral analysis of the same data shows that the dynamics of suspended solids during ordinary conditions is likely to be dictated by the current (Figure 7.16). The spectral variations in the magnitude of the energy density (squared amplitudes) of the current occurs at a frequency corresponding to the frequency of the variation in TSS concentration. The

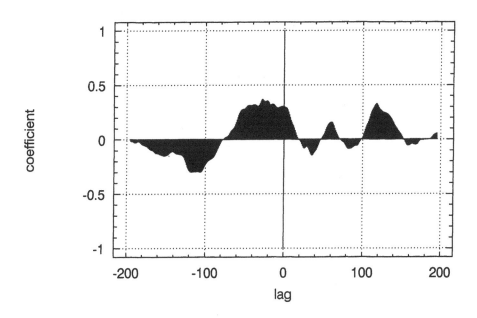

Figure 7.14a. Cross-correlation analysis of current and total suspended solids (TSS) during the period 9-20 December 1993 and 17-26 June 1994. The time lag is a multiple of 4 hours.

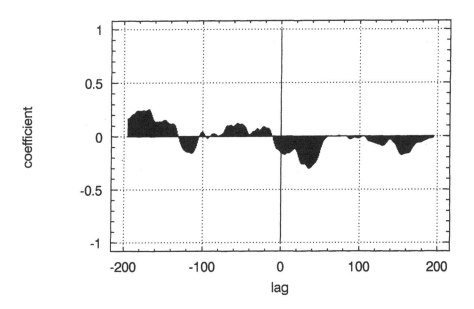

Figure 7.14b. Cross-correlation analysis of TSS and wind during the period 9-20 December 1993 and 17-26 June 1994. The time lag is a multiple of 4 hours.

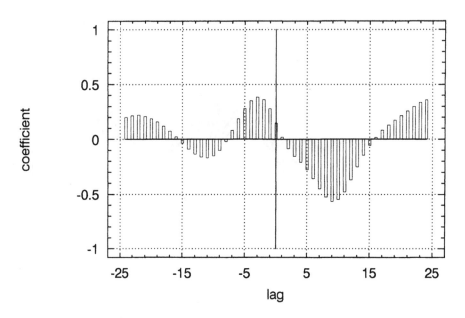

Figure 7.15a. Cross-correlation analysis of current and total suspended solids (TSS) during the period 12-25 March 1994. The time lag is a multiple of 4 hours.

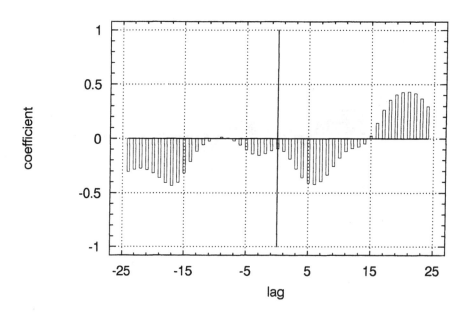

Figure 7.15b. Cross-correlation analysis TSS and wind and TSS during the period 12-25 March 1994. The time lag is a multiple of 4 hours.

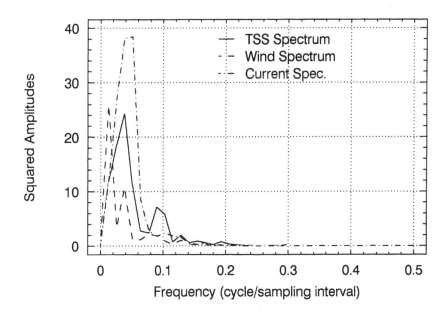

Figure 7.16. Spectral analysis of the TSS, current and wind during the period 12-25 March 1994 in the Bolinao reef.

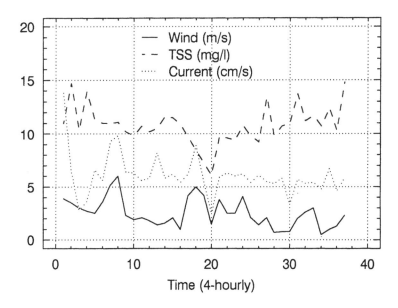

Figure 7.17. Time series of wind, current and TSS concentrations observed in the Bolinao reef during the period from 24-30 June 1994.

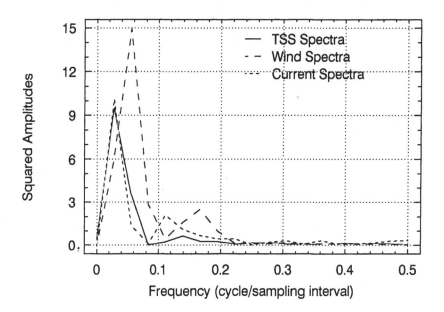

Figure 7.18.   Spectral analysis of the time series in Figure 7.17.

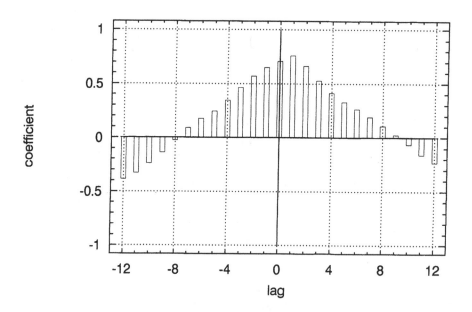

Figure 7.19.   Cross-correlation analysis of current and TSS concentration for a tidal cycle observation (3 April, 1400 H - 4 April 1995, 1300 H) in the Bolinao reef. The time lag is a multiple of 4 hours.

wind spectra show a small positive influence at the frequency of the peak TSS spectra which is attributed to its random nature. In general however, the wind spectra have their peak energy density at a lower frequency which does not correspond to the TSS spectra.

A simultaneous observation of current, wind and TSS concentration during another period of no storm (24-30 June 1994) is shown in Figure 7.17. A spectral analysis of the time series again showed that the current, rather than the wind, is more likely to be the main driving force explaining the dynamics of the observed TSS concentrations (Figure 7.18). A cross-correlation of a tidal cycle observation of TSS concentration and current in the reef during the period 3-4 April 1995 also indicated the significant influence of the current on the dynamics of TSS (Figure 7.19). It is evident that the current effect dominates the dynamics of the TSS concentration during ordinary conditions. However, the effect of the wind in the dynamics of TSS concentration becomes very significant during stormy conditions. The observed temporal variation of the TSS concentrations in the Bolinao reef before and after a strong typhoon (codename 'Katring') is shown in Figure (7.20). The typhoon, with surface winds exceeding 20 m s^{-1}, directly hit the area of investigation in 21 October 1994. The observed TSS concentrations during that period increased from an average of 12 mg l^{-1} to 30 mg l^{-1}. A cross-correlation analysis of the time series of wind and TSS data during this occasion is shown in Figure (7.21).

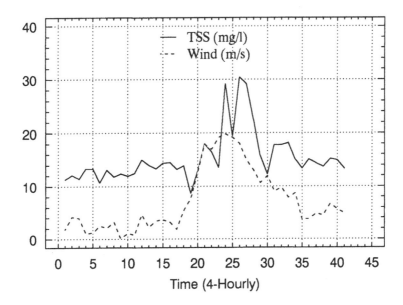

Figure 7.20. Time series of observed wind and TSS in the Bolinao reef during the storm of 20-22 October 1994.

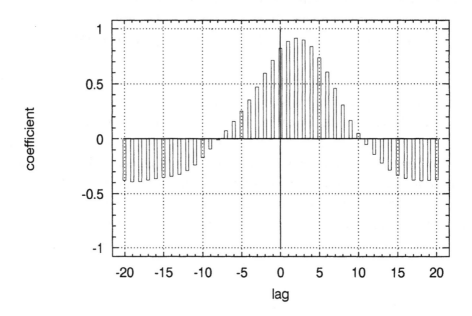

Figure 7.21.   Cross-correlation of the observed wind and TSS in the Bolinao reef during the storm of 20-22 October 1994. The time lag is a multiple of 4 hours.

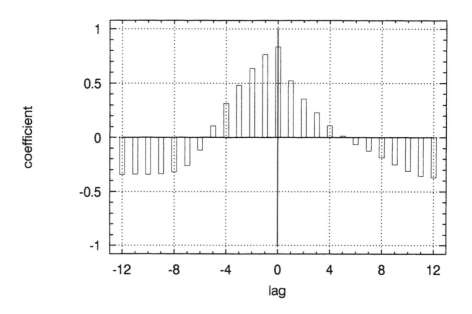

Figure 7.22a.   Cross-correlation analysis of wind and TSS concentration during a tidal cycle (8-9 January 1995) east of the channel. The time lag is a multiple of 1 hour.

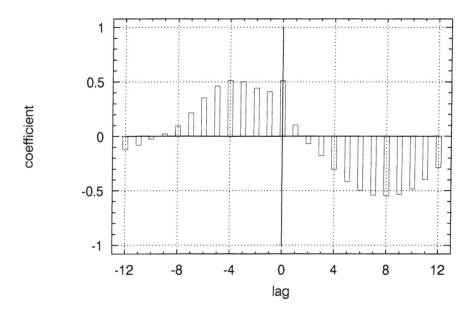

Figure 7.22b. Cross-correlation analysis of current and TSS concentration during a tidal cycle (8-9 January 1995) east of the channel. The time lag is a multiple of 1 hour.

In the channel areas south of Santiago Island, ordinary winds (breeze) can have a significant influence on the TSS dynamics. A tidal cycle observation east of the channel during the period 8-9 January 1995 revealed positive cross-correlation coefficients between the wind and the observed TSS concentrations (Figure 7.22). As shown, the effect of the wind is more significant than the effect of the current in the dynamics of TSS concentration. The predominance of fine sediments (silt-clay range) in this area could be the reason for this observation. It should be noted that these fine sediments can be resuspended already at low bed shear stresses from wind-induced waves.

It can be seen from the foregoing analyses that the short term dynamics (tidal cycle, weekly or monthly) of the TSS concentrations in Cape Bolinao is generally governed by a random component due to the wind and a periodic component due to the (tidal) current. In the reef areas of Cape Bolinao, the wind effect becomes significant only during stormy conditions, while the current effect is significant during ordinary conditions (no storms). The effect of the current can be very significant during spring tides when tidal currents attain their maximum velocity. In the channel areas south of Santiago Island, the wind can have a significant contribution in the TSS dynamics even during ordinary conditions.

On a longer time scale, comparison of separate observations in the Bolinao reef during the southwest and northeast monsoon seasons (with no storms) reveal a weak seasonal variation in the dynamics of TSS concentration (Figure 7.23). The short term perturbations in the TSS concentrations induced by episodic wind events during the southwest monsoon season is not very high. The long-term TSS dynamics for both seasons reveal small fluctuations about a mean concentration which is about equal for both seasons.

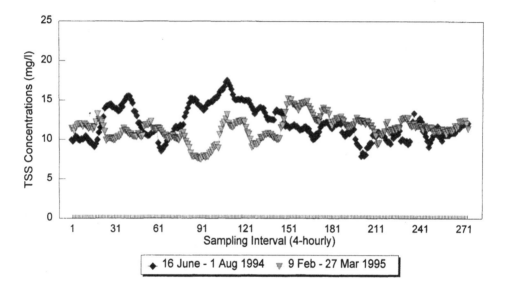

Figure 7.23. Temporal variation of TSS concentrations (mg/l) observed during the southwest (16 June - 1 Aug 1994) and northeast monsoons (9 Feb - 27 Mar 1995) respectively.

Spatially, the weekly observations during the whole period of measurement indicate that sites close to the coast (in the channel) have a relatively higher TSS concentration than sites in the reef north of Santiago Island (Figure 7.24). This can be attributed to the discharge of sediments by creeks and from surface runoff during the southwest monsoon (rainy season). Another reason for the spatial gradient which is true even during the dry season is the possible erosion of unprotected coasts. Eventual sediment transport by tidal and wind-driven currents from the channel to the reef area can occur, but the 'dilution effect' of relatively clear waters north of the reef area (as compared to the channel) can also explain the difference in the observed TSS concentrations.

Measurements of the ash-free dry weight (AFDW) of the observed TSS concentrations during the same period exhibit a relatively high amount of organic matter ($\approx$ 30 %) *in the* suspended solids (Figure 7.25). The enhanced primary productivity in the area is known to contribute to this high amount of organic matter in the TSS concentrations. Generally, the AFDW concentrations exhibit a higher fraction of organic material during calm conditions (no storms). In contrast, during stormy conditions, the fraction of inorganic material is greater than the organic fraction. This implies that resuspension of bottom sediments has occurred.

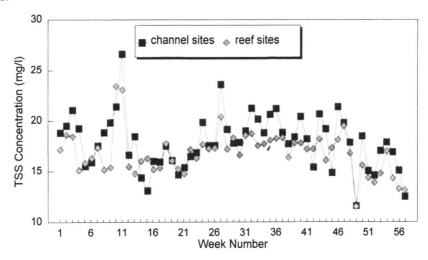

Figure 7.24. Temporal variation (weekly) of the TSS concentrations (mg/l) in Cape Bolinao during the period from 25 April 1994 to 5 June 1995.

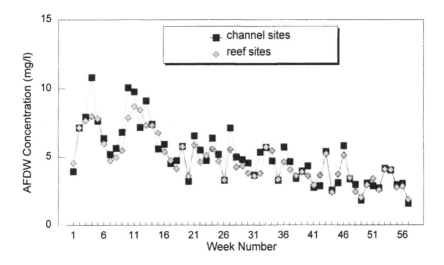

Figure 7.25. Temporal variation of AFDW concentrations (mg/l) in Cape Bolinao during the period 25 April 1994 to 5 June 1995.

7.2.2. Sedimentation Fluxes

The observed sedimentation fluxes in Cape Bolinao revealed a strong spatial difference between the reef and the channel area (Figure 7.26). Measurements in the channel showed consistently higher sedimentation fluxes with an average of 3.2 g m^{-2} hr^{-1} as compared to the reef areas north of Santiago Island with a low average flux of 0.7 g m^{-2} hr^{-1}. The temporal variation of sedimentation fluxes at the reef waters were small during the whole period of observation. In contrast, the channel area showed high fluctuations especially during the southwest monsoon season. While the higher sedimentation fluxes in the channel could be due to the resuspension of bottom sediments, the possible discharge of sediment-laden surface runoff during heavy precipitation could also contribute in the increased fluxes.

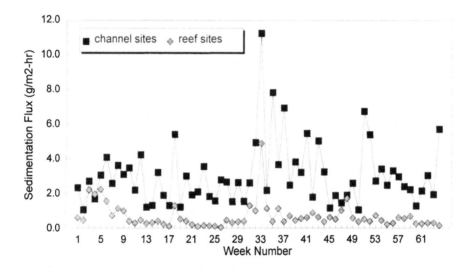

Figure 7.26. Mean sedimentation fluxes at Cape Bolinao during the period from February 1994 to June 1995.

Measurements of the sedimentation fluxes of various sediment fractions (collected by sediment traps) were also undertaken (see Chapter 3). The various size classes of suspended sediments are (operationally) subdivided according to their fall velocity distribution. It was observed that the combined fractions of 'clay' and 'silt' generally showed higher sedimentation fluxes than sand (Figure 7.27). This indicates and further confirms the general observation that most suspended particles during ordinary conditions are in the silt-clay range. During extreme stormy conditions however, there is an increase in the sedimentation

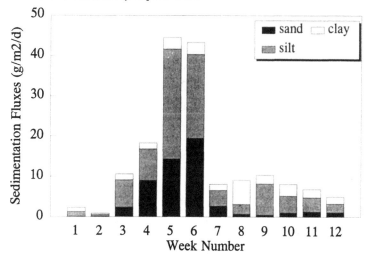

Figure 7.27. Sedimentation fluxes of various sediment fractions in the Bolinao reef observed during the period from March 1994 to April 1995.

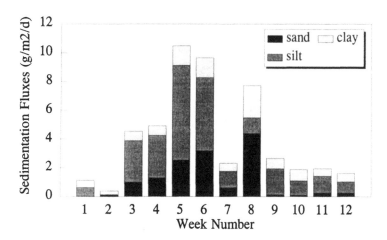

Figure 7.28. AFDW fluxes of various sediment fractions in the Bolinao reef observed during the period from March 1994 to April 1995.

fluxes of sand as seen during the 5th and 6th weeks (10-24 October 1994). This is due to the resuspension and shore erosion of coarser materials during periods of enhanced bed shear stress due to the passage of storms in this period.

Measurements of the associated ash-free dry weight (AFDW) fluxes of these fractions show a variable percentage of organic matter in each size class (Figure 7.28). The organic matter content of the 'sand' and 'clay' fractions is generally smaller than the organic matter content of the 'silt' fraction.

7.2.3. Size Distribution of Bottom Sediments

The bottom sediments off Cape Bolinao are heterogeneous. The bed consist of fractions ranging from clay (< 2 μm) to granules (> 2 mm). There are even coarser sediment fractions from dead corals, as most of the area under investigation is a coral reef area. A typical grain size distribution of the bottom sediments off Cape Bolinao is shown in Figure 7.29.

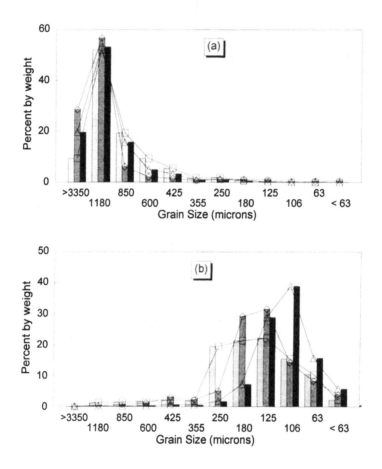

Figure 7.29.   Grain size distribution of bottom sediments off Cape Bolinao. Sites north (south) of Santiago Island are shown in Case a (b).

There is an asymmetrical distribution of bottom sediments, with a dominance of coarser fractions (sand-granules) in the reef area around Santiago Island (case a). The opposite is true south of the island (in the channel) where fine sediments are observed to be dominant (case b). This pattern of sediment size distribution suggests that in the northern areas, the combined action of the waves and currents in those barely protected areas has a major influence on the absence of fine sediments. This is the situation in most sites of the reef which are not protected against waves and strong currents that always act in combination during moderate to strong wind events. The channel area, which is protected by the Santiago Island itself, is less exposed to wave action and strong currents, making it a gross sedimentation zone for discharged and transported sediments, both of allochthonous or autochthonous origin. Such protected areas also mean that the near-bed stress from the combined action of both currents and waves is generally low and below the critical stress for deposition, permitting deposition of even the very fine sediments.

7.2.4. Organic Matter Distribution

The observed distribution pattern of the organic matter content of bottom sediments (averaged for the 3rd and 4th quarter of 1994, and 1st and 2nd quarter of 1995) is shown in Figure (7.30).

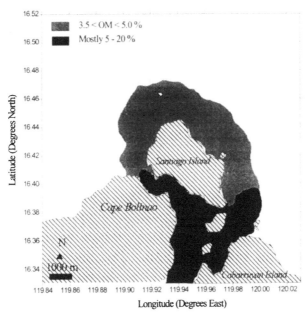

Figure 7.30. Distribution pattern of sediment organic matter content (%) off Cape Bolinao averaged throughout the period of observation.

A north-south gradient in the organic matter distribution is observed. Measured organic content of bottom sediments in the south exceeds 5 % during the whole period of measurement. In some sites in the channel area, the observed values even go beyond 15 %. The content of organic matter decreases in the northward direction, with a percentage below 5 % during the period of measurement. This observation can be correlated to the grain size distribution of fine sediments in the area. The abundance of finer sediments in the south as compared to the north is, on the average, proportional to the amount of organic content of the sediments. The pre-dominance of organic-rich silt fraction in the channel area explains the observed spatial gradient.

It should also be noted that the organic matter content of bottom sediments at the study site is highly correlated with the nutrient content of the sediments. On the average, nitrogen (N) and phosphorus (P) are 2% and 0.5% of the organic matter content respectively at the study site.

7.3.   Light Extinction in Cape Bolinao

The underwater light observations provide information useful for understanding the ecology of the seagrasses, seaweeds, phytoplankton and corals etc. as well as on the processes responsible for the diminution of light which include the dynamics of suspended matter. The available light of interest is confined to the Photosynthetically Available Radiation (PAR) waveband which, as already described in Chapter 5, diminishes in the water column due to the presence of the optically active components such as inanimate suspended particulate, phytoplankton, gilvin, and water itself. The diminution of PAR is expressed through the estimated light extinction coefficient k_d.

The characteristics of the marine environment off Cape Bolinao are different than those of many natural water systems. Of particular significance is the abundance of seagrasses and seaweeds in the area. After decomposition, decay products can contribute much to light extinction by absorption and scattering of PAR. The enhanced phytoplankton growth benefitting from the relatively high amount of organic matter and nutrients in some areas around Cape Bolinao is another factor contributing to light extinction. Furthermore, the presence of relatively high concentrations of suspended sediments (as compared to oceanic water) can make a significant contribution to light extinction. The variation of the light extinction coefficient is primarily governed by the dynamics of the suspended matter. The contribution of water itself in the general extinction of light is another factor. Absorption of PAR by pure water provides a background extinction which becomes dominant when other contributing components are of minor importance.

7.3.1. Light Extinction Dynamics

The estimated light extinction coefficients at Cape Bolinao waters are relatively low with an average k_d of less than 1.0 m^{-1} during the whole period of observation. In the reef area, the estimated light extinction coefficients is generally lower than 0.5 m^{-1}. Higher extinction coefficients, often exceeding 0.5 m^{-1}, were observed in the channel area (Figure 7.31). The relatively higher concentrations of phytoplankton and TSS in the channel as compared to the reef area explains the difference in the observed light extinction coefficients (see 7.3.3).

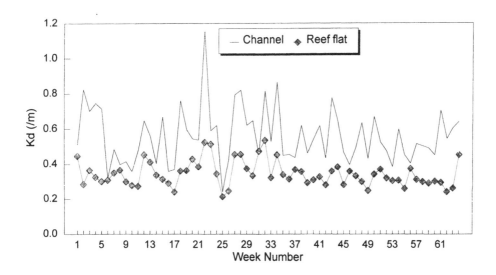

Figure 7.31. Temporal variation of the light extinction coefficient at the Bolinao reef waters during the period from February 1994 to June 1995.

These light extinction coefficients are much lower than those measured in inland waters, but generally higher than in oceanic waters. The presence of the various optically active components (suspended sediments, phytoplankton, gilvin) at various concentrations, all contributed to the observed light extinction coefficients. The lower concentrations of these components in the coastal marine environment as compared to most inland waters explain the smaller extinction coefficients observed.

It is a general observation that the (high frequency) dynamic variation of the light extinction coefficient is attributed to the dynamics of TSS concentrations since the other optically active components (algae and gilvin) are usually not very variable in the short term. A series of continuous observations of TSS and light extinction coefficient obtained during the period from 9-20 December 1993 and 16-26 June 1994 is shown in Figure (7.32).

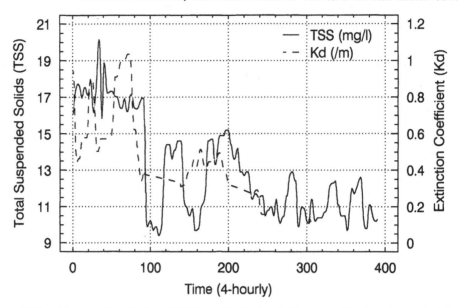

Figure 7.32. Temporal variation of TSS and k_d in the Bolinao reef during the period from 9-20 December 1993 and 16-26 June 1994.

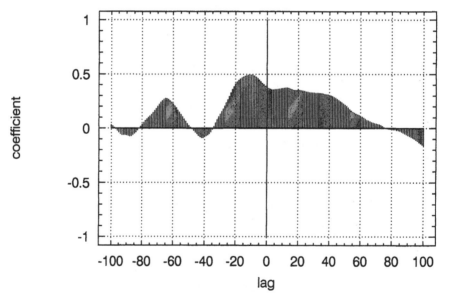

Figure 7.33. Cross-correlation analysis between the time series of TSS and k_d shown in Figure 7.32. The time lag is a multiple of 4 hours.

A cross-correlation analysis between the two time series revealed a positive influence of TSS on the light extinction coefficient (Figure 7.33). A cross-correlation of the observed TSS concentration and light extinction coefficient during the storm of 20-22 October 1994 also showed the positive influence of TSS concentration (Figure 7.34).

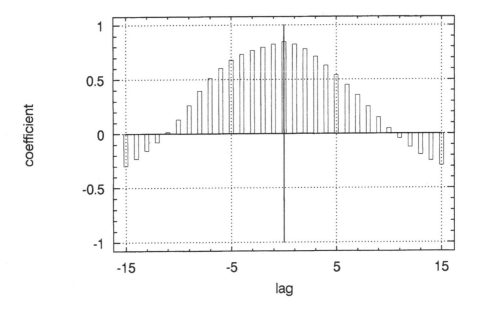

Figure 7.34. Cross-correlation analysis between the time series of TSS and k_d during the storm of 20-22 October 1994 in the Bolinao reef. The time lag is a multiple of 4 hours.

7.3.2. Dependence of Light Extinction on Sediment Size

The light extinction characteristics of various suspended sediments depend on their particle size. The clay particles for example, are known to contribute more to light attenuation than the coarse silt or sand particles. Laboratory measurements on beam attenuation coefficients of the three defined sediment fractions namely sand, silt and clay, were therefore performed. Particles collected by sediment traps were used for this purpose. The measured beam attenuation coefficient in this way is by itself the sum of both absorption and scattering coefficients (see Chapter 6).

7.3.2.1. Beam Attenuation of Sediment Fractions

The fractionation experiments on suspended sediments provided material to execute the spectrophotometric beam measurements for the three sediment sizes assumed (e.g. sand, silt and clay) in the laboratory. Samples from various sites of the Bolinao reef (normally where automatic measurements of TSS and light irradiance are executed) were collected by traps and the three sediment fractions were distinguished according to their respective fall velocities (see Chapter 3 for discussion on the method). The specific beam attenuation coefficient of each fraction is then estimated from the observed beam attenuation coefficients divided by the respective dry weight concentrations of each fraction.

Measurements for the whole photosynthetic range (400 - 700 nm) can not be done in some sites due to the limited capability of the available spectrophotometer. With the Spectronic 20 used, absorbance measurement above the 600 nm wavelength needs a change in the phototube which is not suggested while carrying out the experiments. Figure (7.35) shows the observed spectral variation of the specific beam attenuation coefficients of various sediment fractions.

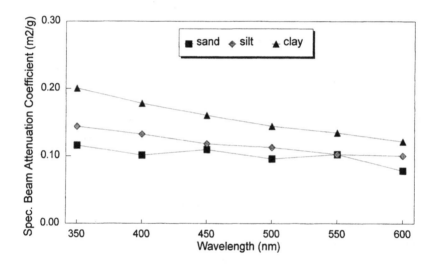

Figure 7.35.   Specific beam attenuation coefficients (averaged) of various sediment size fractions obtained in Cape Bolinao.

The estimated values confirm the inverse relationship between particle size and the beam attenuation coefficients as predicted by theory. As shown in the figure, the specific beam attenuation coefficient decreases in the order clay-silt-sand. Further, there is a general decrease of specific beam attenuation coefficients with increasing wavelength confirming the general observation. This general decrease of the observed specific beam attenuation coefficients with increasing wavelength has also been reported by Kirk (1980) who attributed this to humic materials.

7.3.2.2. Beam Attenuation Measurements of Suspensions

The measurement of beam attenuation coefficients of water samples is difficult to undertake in the coastal sea due to the low concentrations of each of the factors affecting light in the water column. Therefore, several 'turbid sites' were selected for this purpose. In particular, the sites close to the coastal mainland of Cape Bolinao and south of Santiago Island which are known to have relatively higher concentrations of suspended and dissolved matter (gilvin) were chosen. Also, these sites showed relatively higher algal concentrations than the reef sites north of Santiago Island (see 7.3.3).

The specific beam attenuation coefficients from the water samples collected in the designated sites likewise indicate a decrease in the values with increasing wavelengths (see Figure 7.36). The apparent increase with decreasing wavelength is generally attributed to humic substances which may originate from the land and from the decaying seaweeds and seagrasses near the coast.

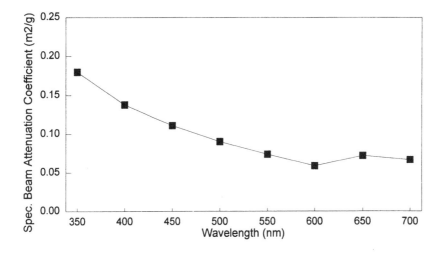

Figure 7.36. Spectral variation of specific beam attenuation coefficients of water samples.

Of particular interest is the observed small shoulder at 650 nm wavelength. The presence of algal particles in the suspensions is thought to be responsible for this. This typical variation manifests the spectral variation of the mean absorption cross-section of phytoplankton component present (see Figure 6.5 in Chapter 6).

7.3.3.   The Contribution of Phytoplankton to Light Extinction

There is an enhanced phytoplankton growth in the channel area of Cape Bolinao. Observed algal concentrations often exceed 1 mg chlorophyll-a m^{-3} in the area (Figure 7.37). Lower concentrations, often below 0.5 mg chlorophyll-a m^{-3}, were observed in the reef. The higher algal concentrations in the channel can be attributed to the relatively higher percentage of organic matter in the area which implies higher amounts of nutrients available for phytoplankton growth (see Section 7.2.4).

A cross-correlation analysis of algal concentrations and the extinction coefficient showed that the contribution of phytoplankton to the light extinction coefficients in the channel area is significant (Figure 7.38). The analysis showed that the algal contribution to light extinction in the area dominates over the contribution of TSS and gilvin. This observation does not apply to the reef area where the algal concentration is low. In the reef, the same cross-correlation analysis showed that the contribution of TSS concentration to the extinction coefficient is dominant.

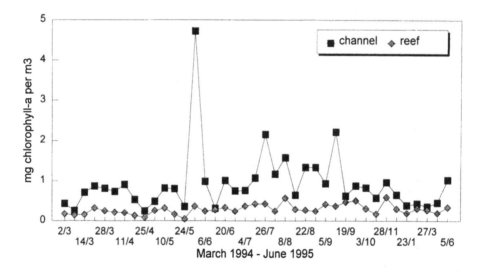

Figure 7.37. Temporal variation of algal concentrations (mg chlorophyll-a m^{-3}) in Cape Bolinao from March 1994 - June 1995.

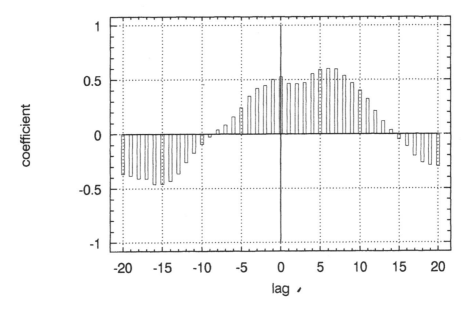

Figure 7.38. Cross-correlation analysis of chlorophyll-a concentration and k_d in the channel area of Cape Bolinao. The time lag is a multiple of 1 week.

To quantify the specific beam attenuation coefficient due to the phytoplankton in the study area, cultured algae at the Bolinao Marine Laboratory were subjected to spectrophotometric analyses. Two diatom species (*Chaetoceros gracilis* and *Isochrysis galbana*) at laboratory concentrations reaching some hundreds of mg chlorophyll-a m^{-3} were used. The measured

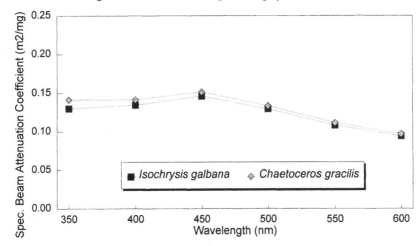

Figure 7.39. Specific beam attenuation coefficient of two diatom species in Cape Bolinao.

specific beam attenuation coefficients of both samples are shown in Figure 7.39. Due to the limited capability of the spectrophotometer used, the phytoplankton absorbance could only be measured up to 600 nm at a scan intervals of 50 nm.

The spectral shape is similar to the general observations made in many natural waters. There is an enhanced absorption at the 450 nm wavelength which decreases towards increasing wavelength (at least until 600 nm). Another peak at 650 nm (lower one) is usually observed near the 650 nm wavelength but could not be measured in our experiments. It is obvious that both diatom species follow the same spectral variation, but *Isochrysis galbana* shows a somewhat lower specific beam attenuation coefficient as compared to *Chaetoceros gracilis* throughout the investigated wavelengths. This could be related to the differences in sizes or shapes of the cells of the two algal species. This typical spectral variation of the specific absorption coefficient is known to correspond to the spectral variation of the mean absorption cross-sections of the phytoplankton (see Section 6.3.2 in Chapter 6).

7.3.4. Gilvin Absorption

The absorption due to gilvin is very variable throughout the observation period at all sites in the study area. Measured absorption coefficients at the 380 nm wavelength vary from a very low (near zero) value to moderately high (above 2 m^{-1}) values. These absorption coefficients are similar to the observed values in some estuarine areas (Figure 7.40). The substantial primary production in the area coupled with land drainage from the mainland and the islands are expected to contribute together to the relatively high absorption coefficients measured. Using Equation (6.5), the spectral variation for three locations were calculated and plotted in the graph below the map. Note the decreasing absorption coefficient with increasing distance from the coast. The observations at A, taken less than 100 m from the coast shows the highest average value, while the farthest (at C) shows the least gilvin absorption coefficient.

7.3.5. Estimated Specific Extinction Coefficients

Rather than the specific beam attenuation coefficient which is just derived from laboratory spectrophotometric analyses, the specific extinction coefficient is the real important quantity to be estimated. Unfortunately, this quantity can not be derived from measurements in a spectrophotometer because in a spectrophotometer, beams of monochromatic light are used and only the beam attenuation due to this monochromatic light is measured. In real marine waters, the whole photosynthetic waveband and its attenuation due to the presence of the different components in the aquatic medium needs to be quantified. Hence, other means of obtaining an estimate of the specific extinction coefficient of each of the various components have to be used. Since measurements were available on the concentrations of TSS (organic

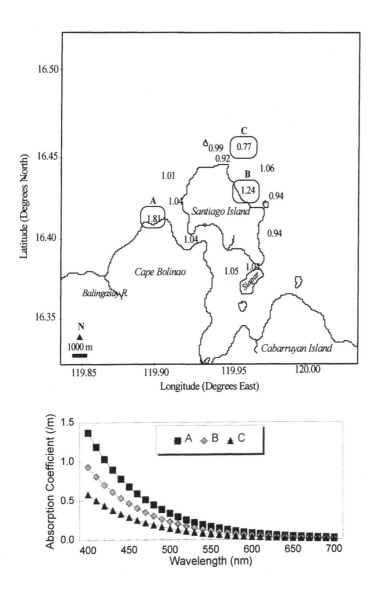

Figure 7.40. Gilvin absorption coefficient measured on several sites at Cape Bolinao. Average values (at 380 nm) throughout the period of observation are presented.

and inorganic), algae, and absorption coefficient of gilvin (at 380 nm), these were used in the analysis. Distinction is made between the organic (detritus) and inorganic (mineral) fraction of TSS since these are known to have different specific extinction coefficients. For the organic part of TSS, the algal concentration is further subtracted. Multiple regression of the observed extinction coefficient and these factors was executed as in Van Duin (1992), Lijklema et al. (1994) and Blom et al. (1994). The results of the regression analysis for the reef and channel sites are shown in Figures (7.41-7.42) with the observed extinction coefficient plotted against the predicted. The estimated specific extinction coefficients, including their standard errors, are provided in Tables (VII.2-VII.3).

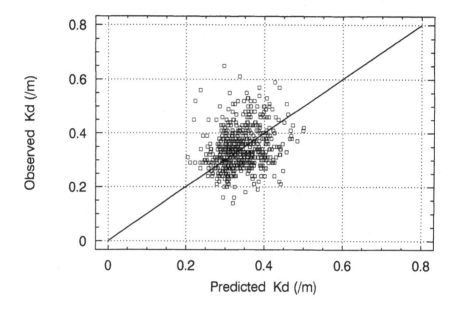

Figure 7.41. Multiple regression analysis of k_d (m^{-1}) as a function of (inorganic and organic) TSS (mg/l), algae (mg chlorophyll-a m^{-3}) and gilvin absorption coefficient (m^{-1}) in the Bolinao reef. The number of observations N = 585.

Component	κ	Estimated	std. error
TSS (organic)	$\kappa_{p,o}$ (m^2 g^{-1})	0.0210	0.0017
TSS (inorganic)	$\kappa_{p,i}$ (m^2 g^{-1})	0.0138	0.0008
Algae	κ_c m^2 mg^{-1} chl-a	0.1485	0.0156
Gilvin	κ_g (-)	0.0253	0.0105

Table VII.2. Estimated specific extinction coefficients in the Bolinao reef.

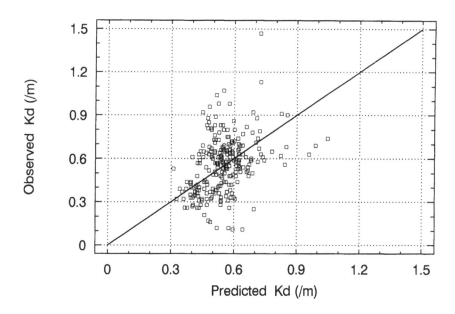

Figure 7.42. Multiple regression analysis of k_d (m^{-1}) as a function of (inorganic and organic) TSS (mg/l), algae (mg chlorophyll-a m^{-3}) and gilvin absorption coefficient (m^{-1}) in the Bolinao channel. The number of observations N = 260.

Component	κ	Estimated	std. error
TSS (organic)	$\kappa_{p,o}$ (m^2 g^{-1})	0.0233	0.0064
TSS (inorganic)	$\kappa_{p,i}$ (m^2 g^{-1})	0.0150	0.0025
Algae	κ_c m^2 mg^{-1} chl-a	0.0736	0.0250
Gilvin	κ_g (-)	0.1409	0.0247

Table VII.3. Estimated specific extinction coefficients in the Bolinao channel.

The specific coefficients obtained for the reef differ slightly with those obtained for the channel. The difference is attributed to the relatively higher extinction coefficients and higher algal concentration measured in the channel (as compared to the reef). For the whole marine system of Cape Bolinao (combined reef and channel), the result of the regression analysis is shown in Figure (7.43). The estimated specific extinction coefficients, including the standard errors, are shown in Table (VII.4). All the components involved were significant at a significance level of ≤ 0.01. However, because of the very low values of the extinction coefficient and concentrations at Cape Bolinao marine waters, it is difficult to obtain a very good fit between the observed and predicted light extinction coefficients. The inherent inaccuracies of measuring the concentrations of the different components and the extinction coefficient itself, contribute to the low correlation. For comparison, literature values according to Buiteveld (1990) are provided in the last column. It should be noted that most literature values were obtained in inland water systems (freshwater). It can be seen that the estimated specific extinction coefficients for TSS (organic and inorganic) in the present study are lower than literature values. The lower values obtained in the present study could be due to the fact that, in the marine environment, suspended sediments tend to flocculate more due to the increased ion concentrations. The resulting increased particle diameters does not only result in an increased settling rate of the flocculated particles but also in a decreased specific scattering and absorption coefficients (Kirk 1983, Blom 1994). Hence, lower specific extinction coefficients could be expected in the marine environment.

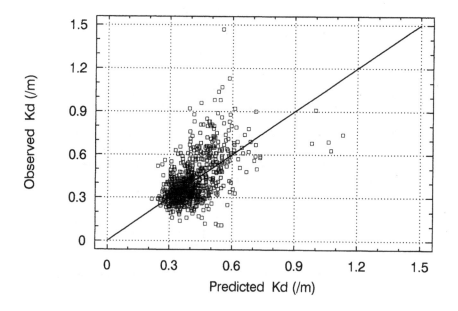

Figure 7.43. Multiple regression analysis of k_d (m^{-1}) as a function of (inorganic and organic) TSS (mg/l), algae (mg chlorophyll-a m^{-3}) and gilvin absorption coefficient (m^{-1}). The number of observations N = 845.

Component	κ	Estimated	std. error	Literature (Buiteveld 1990)
TSS	κ_p (m^2 g^{-1})	0.0134[1] - 0.0147[2]	0.001[1] - 0.0023[2]	0.0253[1] - 0.0490[2]
Algae	κ_c m^2 mg^{-1} chl-a	0.1695	0.0115	0.0209
Gilvin	κ_g (-)	0.0858	0.0123	0.0498

[1] - inorganic TSS

[2] - organic TSS (minus algae)

Table VII.4. Estimated specific extinction coefficients at Cape Bolinao marine waters.

Estimation of specific extinction coefficients of sediment fractions has also been executed in this study. The observed specific beam attenuation coefficients of each of the sediment fractions are used in the estimation procedure. Generally, it can be assumed that there is a linear relationship between the extinction coefficient k_d and the beam attenuation coefficient ϵ (Van Duin 1992);

$$k_d = c_1 + c_2 \epsilon \qquad (7.1)$$

where c_1 and c_2 are constants to be determined. Through linear regression of the observed extinction coefficients and the measured beam attenuation coefficients, the constants were obtained. The specific extinction coefficients of the sediment fractions can be determined by correcting the measured specific beam attenuation coefficient with the derived constant, c_2. The increment c_1 is generally neglected since a zero specific beam attenuation coefficient means a zero specific extinction coefficient. The estimated partial specific extinction coefficient of the three sediment fractions can be seen in Table VII.5. The estimation yielded low correlation coefficient ($r < 0.5$) which is due to the limited number of data collected.

fraction	κ (m^2 g^{-1})
"clay"	0.014
"silt"	0.010
"sand"	0.009

Table VII.5. Estimated specific extinction coefficients of the different sediment fractions.

It should be noted that measurements of the beam attenuation coefficient of the sediment fractions were performed at a later stage of the research only. Nevertheless, the estimated specific extinction coefficients for the three sediment fractions confirm the general observation that finer sediment fractions contribute more than coarser sediments per unit weight material.

Chapter 8

Sensitivity Analysis, Calibration and Validation of the Numerical Models

The results of the modelling studies on the hydrodynamics, sediment transport and light extinction in Cape Bolinao are presented in this chapter. Most of the mathematical equations solved in the modelling exercise have been presented in the previous chapters. It should be noted that there are inherent uncertainties in some of these equations. The parameterization of some processes represented by these equations is only an approximation of the actual physics involved. In the hydrodynamic model for example, the bulk aerodynamic formula for estimating surface stress from the wind involves the use of a drag coefficient which has no unique value. The same is true for the bottom drag coefficient in the conventional bottom friction formula. Even more parameters are involved in the sediment transport model which define the fluxes of resuspension and sedimentation. From estimating the significant wave characteristics to estimating the wave-related bed shear stress, several parameters are needed which are not known a priori. The same is true with the current-related stress parameters. The absence of actual field measurements of these physical processes precludes the determination of universal parameters. In effect, literature values of some parameters differ greatly, sometimes by two orders of magnitude. Concerning the light extinction model, there are also uncertainties in the specific extinction coefficients obtained through multiple regression as the output of such analysis is strongly dependent on the quality of available data. Here, measurement errors which could not be avoided, all contribute to uncertainties in the estimated specific extinction coefficients.

In general, the values of the relevant parameters are dependent on the measurement site and the existing conditions in the particular area of study, i.e. they are site-specific. Therefore, relying only on literature values may not give a good description of the investigated processes at a particular site. In order to arrive at a reliable predictive model for Cape Bolinao and the Lingayen Gulf, these parameters which are poorly known should therefore be adjusted in order to get a good fit between the simulated and observed data, i.e. the model has to be tuned or calibrated. Due to the complexity of the hydrodynamic and sediment transport models, a 'trial and error' calibration procedure was used after which an optimum set of parameter values was derived. The calibration procedure, encompassing model sensitivity analysis, is addressed in Sections (8.1-8.3). Validation of the numerical models of the hydrodynamics, sediment transport and light extinction in Cape Bolinao is presented in Section (8.4). The data obtained during the observational period was used for this purpose.

8.1. The Hydrodynamic Model

8.1.1. Sensitivity to Advection of Momentum

The advection of momentum, represented by the first-order non-linear terms in the hydrodynamic model is usually neglected in many coastal models describing the dynamics of long-wave induced currents due to the wind and tide. The reason for the exclusion of the advective terms are that, their contribution to the magnitudes of the currents is small, and their inclusion in the model greatly increases the computational time. Furthermore, the difficulty in obtaining accurate numerical solutions for the non-linear convective terms remains a problem.

In general however, the advective terms are important in shallow areas (Flather and Heaps 1975, Xian Xao 1993). Especially in areas near the coast with abrupt bathymetric change, the associated non-linear terms are not negligible, hence, they are important when modelling the general circulation in the coastal zone. Another reason for including advection is that, it has a cumulative influence on the flows which means that it is important when modelling the long-term evolution of the flow velocities using a time-dependent model. It can be shown through mathematical analysis that it plays an important role in the dissipation process of vorticity (circulation per unit area). As in (Xian Xou 1993), assuming a constant wind stress and linear bottom friction, the (potential) vorticity is given by

$$\eta \frac{d}{dt}\left[\frac{\xi + f}{h}\right] + \frac{kU}{h}\frac{\xi + f}{h} = \frac{kU}{h}\frac{f}{h} + \frac{1}{h^3}\left[\left(kUv - \frac{\tau_{sy}}{\rho}\right)\frac{\partial h}{\partial x}\right.$$
$$\left.\left(kUu - \frac{\tau_{sx}}{\rho}\right)\frac{\partial h}{\partial y}\right] \qquad (8.1)$$

where ξ (= $\partial v/\partial x$ - $\partial u/\partial y$) is the relative vorticity, f is the Coriolis parameter, U is a reference (characteristic) flow velocity, k is the bottom friction coefficient, and η is a coefficient which determines whether advection is considered ($\eta = 1$) or neglected ($\eta = 0$).

Equation (8.1) describes the process of vorticity dissipation through the bottom frictional resistance with a typical timescale for decay T_d and a lengthscale for dissipation L_d given by

$$T_d = \frac{\eta h}{k U} \quad , \quad L_d = \frac{\eta h}{k} \tag{8.2}$$

It can be seen from Equation (8.2) that the vorticity is instantaneously dissipated ($L_d = 0$) if the advective terms are neglected ($\eta = 0$) (Xian Xou 1993). Hence, in modelling the temporal evolution of flow velocities, advection becomes important.

The use of the expanded advective terms in the present hydrodynamic model has to be emphasized as well (see Chapter 4). This 'modified advection' has been derived by Koutitas (1988) by assuming a quadratic profile of current in the vertical. The advantage of using the expanded advection terms is that, non-uniformity of the current profile in the vertical is taken into account. This can be seen as a correction imposed on advection which is important in simulating wind-driven flows in the coastal zone, as wind-driven flows are highly non-uniform. It was also shown by Koutitas (1988) that this modification does not only result in a faster convergence of the numerical solution but also to a more reasonable simulation of the wind-driven surface elevation in the coastal zone. Hence, in the present application, this modified model is adopted.

Numerical simulations of the wind and tide-driven flows in Cape Bolinao with and without the expanded advective terms are presented in Figures (8.1-8.2). The model is run with a uniform wind speed of 3 m s^{-1} from the north-northwest direction (annual mean) in conjunction with the tidal forcing described by Equation (4.44). The surface drag coefficient in both exercises (with and without advection) is assumed constant ($c_d = 0.0035$). As the model includes the tidal forcing, results for both flood and ebb tide are shown for comparison. These results show that the contribution of the advective terms to the magnitude of the currents is small as the magnitudes of the flow velocities, with and without advection, are very similar. This is true during both flood and ebb tide. In his quasi-three dimensional model, Xian Xou (1993) also noted the small contribution of advection to the magnitudes of the simulated currents.

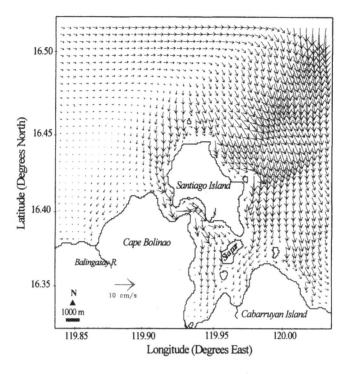

Figure 8.1a. Simulated flow pattern in Cape Bolinao at flood tide (*standard condition*).

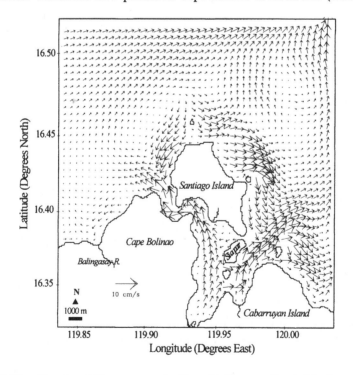

Figure 8.1b. Simulated flow pattern in Cape Bolinao at flood tide (*standard condition*).

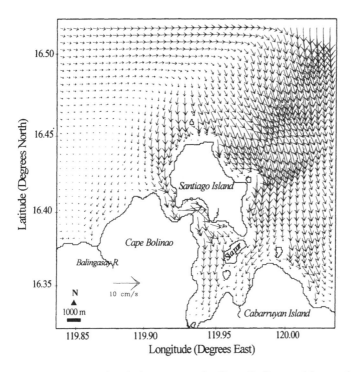

Figure 8.2a. Simulated circulation pattern in Cape Bolinao without advection (flood tide).

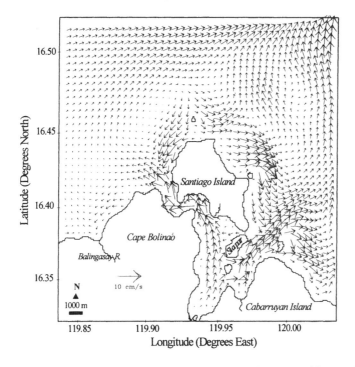

Figure 8.2b. Simulated circulation pattern in Cape Bolinao without advection (ebb tide).

Concerning the effect of advection on the sea surface elevation, Figure (8.3) shows the simulated time series of water level (tide) near the coast of Cape Bolinao during a storm event. It can be seen that, before the storm ($t < 116$ h), advection has a negligible effect on the sea surface elevation. However, the effect of the advective terms become noticeable during the storm (116 h $< t <$ 160 h). Here, surface winds approaching 20 m s^{-1} were experienced in the area. As shown, the surface elevation during the storm is higher when advection is included. The difference in the sea surface elevation is expected to increase when the storm winds are stronger. This is actually one of the obvious advantages of the modified model. It results in higher free surface gradients balancing both the free surface shear and the bed shear (in the same direction) and consequently to a reasonably higher sea surface elevation along the coast (Koutitas 1988).

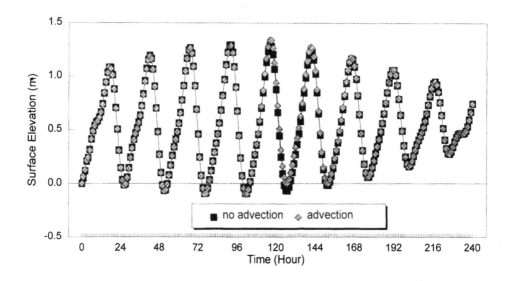

Figure 8.3. Simulated sea surface elevation in Cape Bolinao with and without advection.

Generally, however, the results suggest the relatively small influence of advection in the hydrodynamics of Cape Bolinao. This is generally attributed to the (still) large grid distance used in the numerical model ($\Delta s = 500$ m). The grid distance should be much smaller than the dissipation length L_d for advection to become significant. With depths of about 10 m in the main area of interest (the Bolinao reef and surrounding) and taking $k = 0.01$ as a representative value for the bottom friction coefficient, $L_d = 1000$ m which is still comparable to Δs. Nevertheless, the advective terms are retained in the present modelling application to ensure conservation of momentum and faster convergence of the numerical solutions.

8.1.2. The Effect of Momentum Diffusion

To investigate the effect of momentum diffusion, the hydrodynamic model is run with and without the diffusion terms (instead of changing the values of the diffusion coefficients). A fixed value for the diffusion coefficient is used ($A_h = 10$ m^2 s^{-1}). The same wind and tide forcing as applied in Section (8.1.1) are used in the sensitivity analysis. Results of the simulation during flood and ebb tide are presented in Figure (8.4). The results at flood tide show the distinct formation of the weak anticyclonic circulations northwest of Santiago Island and Cape Bolinao. Comparing these with the standard condition (Figure 8.1a), the velocity gradients near the center of the gyres appear stronger without the diffusion terms. This is due to the fact that diffusion tends to smooth out gradients and neglecting it will certainly result in more pronounced velocity gradients. The effect of diffusion on the sea surface elevation can also be seen in Figure (8.5).

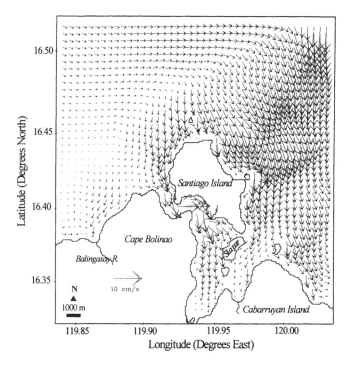

Figure 8.4a. Simulated circulation pattern in Cape Bolinao without diffusion (flood tide).

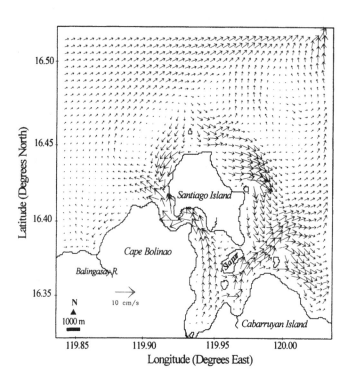

Figure 8.4b. Simulated circulation pattern in Cape Bolinao without diffusion (ebb tide).

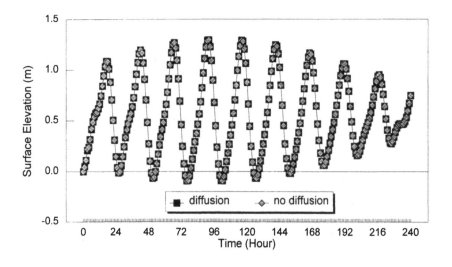

Figure 8.5. Simulated sea surface elevation in Cape Bolinao with and without diffusion.

As shown, diffusion has a negligible effect on the sea surface elevation, as both results (with and without diffusion) correspond very well. Similar to advection, the influence of diffusion in the general circulation of Cape Bolinao appears to be small. However, their retention in the model and the use of constant diffusion coefficients are recommended as the induced physical/numerical diffusion smooths out numerically induced perturbations in the velocity field (Koutitas 1988).

8.1.3. Wind-driven Currents - Surface Stress Effect

Investigation on the effect of surface stress on the wind-driven circulation of Cape Bolinao is mainly confined to changing the value of the drag coefficient c_d in the surface stress formulation. In general, the drag coefficient is weakly dependent on the magnitude of the wind speed (see Wu's drag formulation in Equation 4.12). To examine its effect, the model is run using Equation 4.12 (giving $c_d \approx 1 \times 10^{-3}$) and a higher constant drag coefficient value ($c_d = 0.0035$). For both simulations, the mean annual wind speed (3 m s^{-1}) and direction (NNW) is used. With both advection and diffusion included in the hydrodynamic model, the results of the simulation are shown in Figure (8.6).

A general feature of the simulated wind-driven circulation (shown by both results) is the formation of the weak anti-cyclonic circulations before the reef slope northwest of Santiago Island. These can be seen also in the general circulation in Figures (8.1-8.2). Apparently, they are wind-driven. The clockwise deflection in the current directions in this area is the result of flow modification by the abrupt bathymetric change, the irregular coastal geometry of Cape Bolinao and the prevailing wind direction. It should be noted that at the reef slope, water depths increase abruptly. Also, mass transport through the Bolinao channel is partly obstructed by its small opening. The observed current pattern is also the result of the prevailing wind direction (NNW). It is shown in Section 8.4, that this general circulation pattern does not appear during the southwest monsoon season.

Comparing Figure (8.6a) with Figure (8.6b), it can be seen that using Wu's formulation results in smaller magnitudes of simulated depth-mean currents in Cape Bolinao (see scale). Table VIII.1 also shows some computed depth-mean currents using Wu's drag coefficient (0.001) and a higher constant drag coefficient of 0.0035. In three locations in the Bolinao reef, a great difference in the current magnitudes ($\approx$ 51 %) was noted. When the actual observations on surface current are used for comparison, it is evident that the surface currents are underestimated using Wu's drag formulation. From model calculations, the surface currents can be estimated from the depth-mean currents using the equation

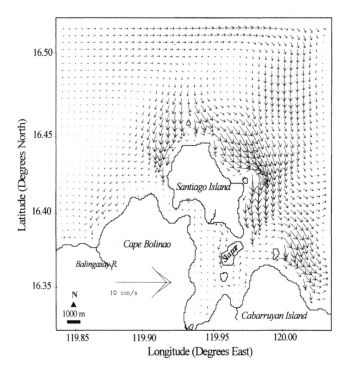

Figure 8.6a.   Simulated circulation pattern in Cape Bolinao using $c_d = 0.001$.

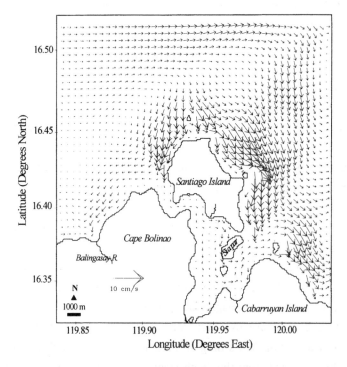

Figure 8.6b.   Simulated circulation pattern in Cape Bolinao using $c_d = 0.0035$.

$$u_{surf} = 1.5u + a_x/4 \quad , \quad v_{surf} = 1.5v + a_y/4 \qquad (8.3)$$

where u_{surf} and v_{surf} are the magnitudes of the x and y-components of surface currents

c_d	Loc 1	Loc 2	Loc 3
0.0010	2.33 cm s^{-1}	1.92 cm s^{-1}	2.16 cm s^{-1}
0.0035	4.54 cm s^{-1}	3.76 cm s^{-1}	4.20 cm s^{-1}
% change	51.32 %	51.06 %	51.42 %

Table VIII.1. Computed depth-mean currents in three locations in the Bolinao reef.

In accordance with Equation (4.22), the stress-related terms a_x and a_y add up a flow speed of less than 2.5 cm s^{-1} to the surface current (with a wind speed of 3 m s^{-1}). With simulated u and v of less than 5 cm s^{-1} using Wu's drag formulation, the surface currents are generally less than 10 cm s^{-1}. From the result of the model simulation with a higher value of c_d, it can be shown using Equation (8.3) that the simulated magnitudes of current become more reasonable when compared with observations. This is shown in the model validation in Section 8.4.

8.1.4. Tide-Driven Circulation

The contribution of the tide in the general circulation pattern of Cape Bolinao was examined separately. Investigation of the individual contribution of the tide can be done by setting the wind to zero. Because of the non-linear interaction that is expected between the wind and the tide, it is not proper to determine the tide-driven currents by subtracting the wind-driven current from the general model simulation with wind and tide forcing as input. For simulating the tide-driven currents, the present non-linear model has to be modified. It should be noted that this model does not apply for pure tide-driven flows since it was originally derived for wind-driven circulation with wind stress specified as boundary condition in the derivation. In particular, the advective terms have to be reduced to their original forms (Equations 4.1-4.2) when the wind effect is neglected. An important consequence of this modification is that, the finite difference forms of the advective terms in the general equations used (Equations 4.17-4.18) have to be changed for reasons of numerical stability. In particular, the terms $u\partial u/\partial x$ and $v\partial v/\partial y$ are given by (Koutitas 1988)

$$u\frac{\partial u}{\partial x} = \frac{1}{2}\frac{\partial u^2}{\partial x} = \frac{1}{8\Delta s}\left[\left(u_{i+1,j}^n + u_{i,j}^n\right)^2 - \left(u_{i,j}^n + u_{i-1,j}^n\right)^2\right]$$
$$v\frac{\partial v}{\partial y} = \frac{1}{2}\frac{\partial v^2}{\partial y} = \frac{1}{8\Delta s}\left[\left(v_{i,j+1}^n + v_{i,j}^n\right)^2 - \left(v_{i,j}^n + v_{i,j-1}^n\right)^2\right] \qquad (8.4)$$

The other advective terms (i.e. $v\partial u/\partial y$ and $u\partial v/\partial x$) do not include the correction and stress-related terms and are computed by central differences (no interpolation as in Equation 8.4). Furthermore, the conventional quadratic bottom friction (Equation 4.13) has to be applied for pure tide-driven flows since the modified friction relationships (Equation 4.24) result in a zero bottom friction when the wind speed is zero. Due to the absence of frictional dissipation, this tends to give unstable numerical solutions. With the use of the conventional quadratic friction law, the solutions become stable.

The results of the simulation for pure tide-driven flows for both flood and ebb tides are shown in Figure (8.7). As in Koutitas (1988), the bottom drag coefficient k is assigned a value of 0.01 which corresponds to a Chezy coefficient of 31.3 $m^{1/2}$ s^{-1}. The results for both

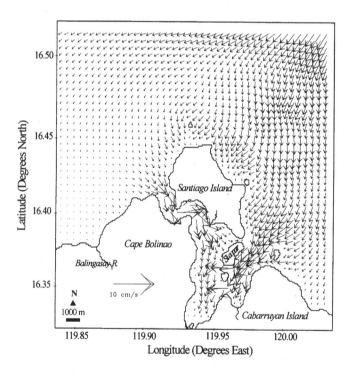

Figure 8.7a. Simulated circulation pattern in Cape Bolinao for pure tide forcing (flood tide).

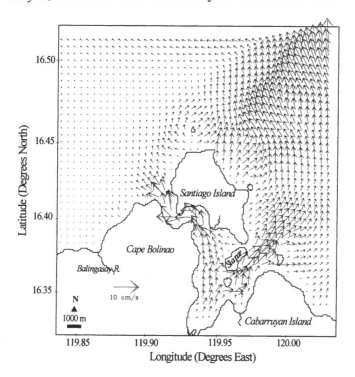

Figure 8.7b. Simulated circulation pattern in Cape Bolinao for pure tide forcing (ebb tide).

flood and ebb tide do not show the anti-cyclonic circulation. At flood tide, there is a general southward flow, while a northward flow appears at ebb tide. For both tidal regimes, the current directions are strongly influenced by the irregular topography of Cape Bolinao. The flood currents tend to penetrate inside the channel bordering Santiago Island and Cape Bolinao (Figures 8.7a). The opposite is observed during ebbing (Figure 8.7b) with tidal currents going out of the channel. In general, the simulated ebb currents ($\approx$ 10 cm s^{-1}) are stronger than the flood currents ($<$ 10 cm s^{-1}), which is in accordance with observations presented in Chapter 7.

8.2. The Sediment Transport Model

8.2.1. The Resuspension Flux

In the sediment transport model, the resuspension flux is probably the most difficult to estimate since no field measurements of this variable are available. This is also true for the wave and current-stresses responsible for resuspension. Investigation on the effect of the resuspension flux on the dynamics of suspended sediment concentration and transport is a broad exercise involving many parameters, e.g. resuspension constant, critical stress, wave and current friction factors, and many other relevant parameters. For the purpose of investigating the effect of the different parameters in the resuspension flux, the formulation (Equation 5.31) according to Parchure and Mehta (1985) is used. The reason for using this resuspension formulation in the sediment transport model and basis of a preliminary calibration is that it gives good estimates of the suspended sediment concentrations when compared with observations. The sensitivity analysis presented here concerns the influence of the resuspension constant, α-parameter, and critical shear stress. The relative influence of current and waves is also investigated. The following sections present the results of the modelling exercise.

8.2.1.1. Effect of the Resuspension Constant

The resuspension constant k_r (also called the floc erosion rate) shows a strong influence on the dynamics of TSS concentration. Reported values of k_r for fine cohesive sediments range from 6.6×10^{-8} to 5.3×10^{-6} kg m^{-2} s^{-1} (Parchure and Mehta 1985). In a series of numerical simulations conducted in the present study, the higher value appear more reasonable for Cape Bolinao sediments. For the clay, silt and sand fractions assumed in the model, k_r values of, respectively, 2.0×10^{-7}, 3.0×10^{-6} and 5.5×10^{-6} kg m^{-2} s^{-1} were found to be reasonable estimates. To show the strong influence of the resuspension constant on the TSS concentrations, a two-fold increase and decrease of these k_r values were adopted. A mean annual wind speed of 3 m s^{-1} (NNW mean annual direction) in conjunction with the tide was used to force to hydrodynamic model. The simulated distribution patterns of the total suspended sediment concentrations in Cape Bolinao for this modelling exercise are shown in Figure (8.8a,b). It can be seen that, a two-fold increase in the value of k_r results in unreasonably high TSS concentrations (Figure 8.8a). The opposite is observed for a two-fold decrease in k_r, giving an underestimation of observed TSS concentrations.

8.2.1.2. Effect of the α-parameter

The α-parameter in the resuspension model is known to be inversely proportional to the absolute temperature. Typical values reported in literature range from 4.2 to 25.6 m/N$^{1/2}$ (Parchure and Mehta 1985). However, it is shown in a series of numerical simulations that these values are too high for Cape Bolinao. The strong dependence of the simulated sediment concentrations on the α-parameter is shown in Figure (8.9). In the experiments,

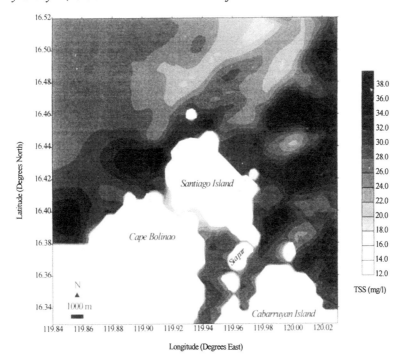

Figure 8.8a. Simulated TSS distribution pattern in Cape Bolinao with a two-fold increase in the resuspension constant k_r.

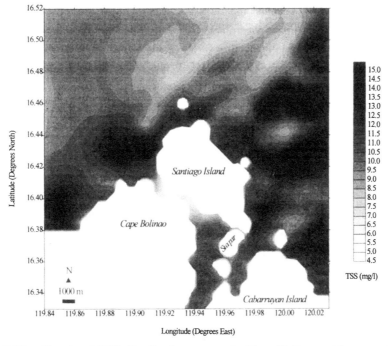

Figure 8.8b. Simulated TSS distribution pattern in Cape Bolinao with a two-fold decrease in the resuspension constant k_r.

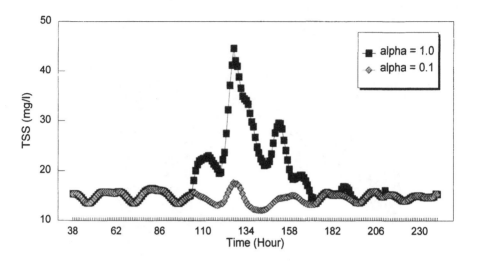

Figure 8.9.   Simulated TSS concentration (g m^{-3}) with $\alpha = 1.0$ and $\alpha = 0.1$.

two values of α were used, i.e. case a, $\alpha = 1.0$ and case b, $\alpha = 0.1$. The same hydrodynamic forcing used in Section 8.2.1.1 is applied in this exercise. For $\alpha = 1$, the simulated sediment concentrations are considerably higher than observed TSS concentrations. For $\alpha = 0.1$, the simulated concentrations are, in contrast, much lower. When compared with observations in Cape Bolinao, a parameter value in the range ($0.1 < \alpha < 1$) appears to be representative for Cape Bolinao. In a series of numerical experiments, it was found that $\alpha = 0.5$ is reasonable for Cape Bolinao sediments. The simulation results using this value of α are shown in the model validation (Section 8.4).

8.2.1.3. Effect of the Critical Shear Stress τ_{cr}

The critical shear stress for resuspension is dependent on the sediment sizes and densities. For the three sediment fractions assumed in the present study, the critical shear stresses were estimated using the Shields approach (see Chapter 5). Using representative sizes for the different fractions, the estimated critical shear stresses were, respectively, 0.14, 0.15, and 0.19 N m^{-2} for the 'clay', 'silt' and 'sand' fractions. For a sensitivity analysis on the critical shear, an order of magnitude increase and decrease of these τ_{cr} values were used. Using the same hydrodynamic forcing as in the previous sections, the influence of the critical shear stress τ_{cr} on the total suspended sediment concentrations is shown in Figures (8.10-8.11). It can be seen that the critical shear stress does not have a very strong influence on the TSS concentrations (as compared with k_r and α). An order of magnitude increase and decrease of τ_{cr} results in just a two-fold increase of the TSS concentration (Figure 8.10). This observation is also evident in the TSS distribution patterns shown in Figures (8.11a-8.11b).

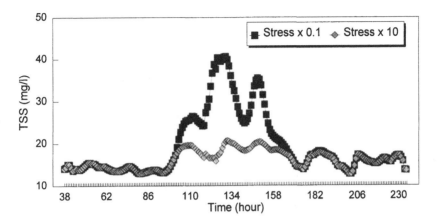

Figure 8.10. Simulated TSS concentration (g m⁻³) with an order of magnitude increase and decrease in the critical shear stress τ_{cr}. The α-parameter in the resuspension flux is set at 0.5 m/N$^{1/2}$.

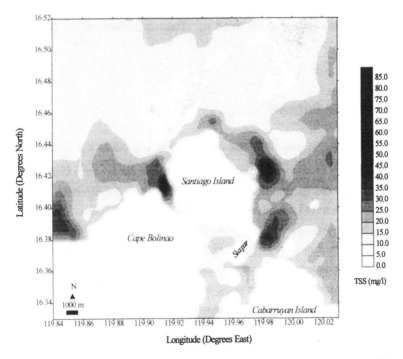

Figure 8.11a. Simulated TSS distribution pattern in Cape Bolinao with an order of magnitude increase in the critical shear stress τ_{cr}. The α-parameter in the resuspension flux is set at 0.5 m/N$^{1/2}$.

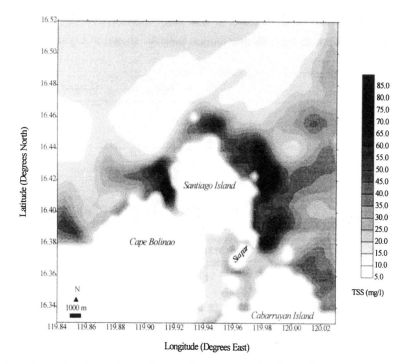

Figure 8.11b.   Simulated TSS distribution pattern in Cape Bolinao with an order of magnitude decrease in the critical shear stress τ_{cr}. The α-parameter in the resuspension flux is set at $0.5 \ m/N^{1/2}$.

8.2.1.4. Wave-Current Influence

To get an estimate of the relative influence of waves and current in the dynamics of TSS in Cape Bolinao, numerical experiments were conducted separately for both factors. The results of these exercises are shown in Figures (8.12-8.13). Both simulations were done with stormy wind conditions to get an appreciable effect from both factors. In Figure (8.12), it is shown that the wave influence dominates the TSS concentrations in the Bolinao reef. This result is taken north of Santiago Island. However, as shown in Figures (8.13a,b), the current-induced TSS concentrations are of comparable magnitudes to the wave-induced concentrations. The areal maps further show that the wind-induced TSS concentrations are even higher in some sites. It is evident that during enhanced wind speeds, there is a significant influence of both waves and current on the total suspended sediment concentrations. Both factors have to be considered therefore for an accurate modelling of the suspended sediment transport in Cape Bolinao.

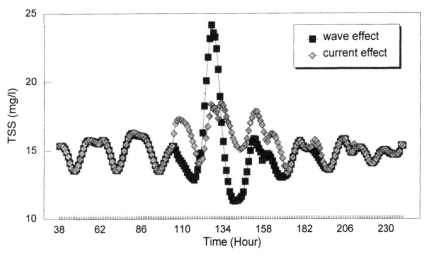

Figure 8.12. Simulated TSS concentration (g m^{-3}) due to waves and a current.

8.2.2. The Sedimentation Flux

In a series of numerical tests involving several formulae for the sedimentation flux ϕ_s, it was found that using the sedimentation model $\phi_s = w_s(c - c_o)$ gives a better estimate of the suspended sediment concentrations in Cape Bolinao than other formulations. The formula

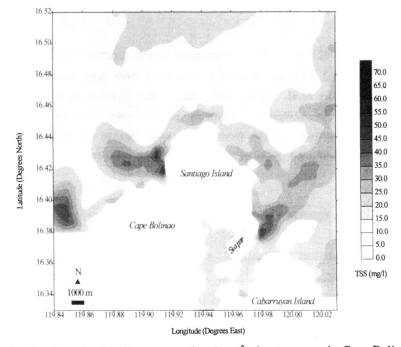

Figure 8.13a. Simulated TSS concentration (g m^{-3}) due to waves in Cape Bolinao. The α-parameter in the resuspension flux is set at 0.5 m/N$^{1/2}$.

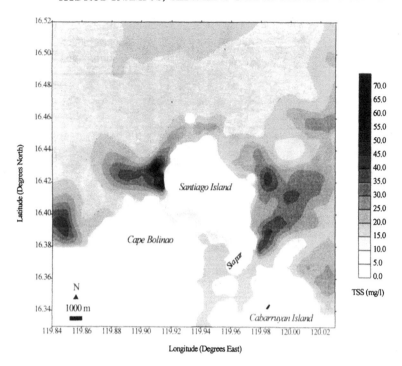

Figure 8.13b. Simulated TSS concentration (g m⁻³) due to currents in Cape Bolinao. The α-parameter in the resuspension flux is set at 0.5 m/N$^{1/2}$.

according to Krone (1962) where the sedimentation flux is function of the probability of deposition ($\phi_s = pw_sc_b$) was not successful because it gave high suspended sediment concentrations. Lijklema et al. (1994) also noted similar observation where there is an unlikely accumulation of suspended sediments in the water column using Krone's formulation. The sensitivity of the model predictions due to the sedimentation flux formulation was then examined, particularly for the settling velocity w_s.

8.2.2.1. Effect of the Settling Velocity w_s

The estimation of the settling velocities, w_s, of the sediment fractions was done using the Stoke's law as a first approximation. From Stoke's law, the estimated settling velocities from clay to sand fractions are on the order of 10^{-4} to 10^{-3} m s⁻¹. After a series of numerical experiments, settling velocities of 1.0×10^{-3}, 5.5×10^{-4} and 3.5×10^{-4} m s⁻¹ were found to be representative for the 'sand', 'silt' and 'clay' fractions respectively. In general, there is a strong influence of the settling velocity on the simulated suspended sediment concentrations (Figure 8.14). For example, a two-fold decrease in the settling velocities ($w_s/2$) results in an unrealistic increase in the sediment concentrations.

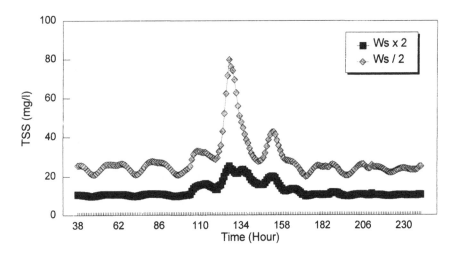

Figure 8.14. Simulated TSS concentration (g m^{-3}) with a two-fold increase and decrease in the sediment settling velocity w_s.

On the other hand, a two-fold increase in the settling velocities (w_s x 2), result in (very) low TSS concentrations. As shown, the total suspended sediment concentrations are always below 20 g m^{-3} with an order of magnitude increase in w_s. This is not representative of observed TSS concentrations in Cape Bolinao (see Chapter 7).

8.3. The Light Extinction Model

The light extinction model which was developed and applied in this study follows the general linear relationship discussed in Chapter 6. The result of the regression analysis in the previous chapter provided the input specific extinction coefficients for the light extinction model. The predictions of the sediment transport model for three sediment fractions are then used in modelling the light field in Cape Bolinao. The concentration of algae and the absorption coefficient of gilvin were assumed constant since these two components do not vary significantly in the short term. Based on observations, a mean algal concentration of 0.3 mg chlorophyll-a m^{-3} and a mean gilvin absorption coefficient of 0.8 m^{-1} (at 380 nm) were used. Furthermore, a distinction between organic and inorganic fractions in the total sediment concentrations was made. In accordance with observed proportions of AFDW and TSS concentrations in Cape Bolinao, the organic fraction was about 30% of the total sediment concentration. The inorganic fraction was then determined from the difference of the total sediment concentration (combined sand, silt and clay) and the organic fraction.

8.3.1. The Specific Extinction Coefficients

Two sets of specific extinction coefficients were used in the sensitivity analysis and calibration of the light extinction model. The first set comprised the coefficients obtained from the present study, while the second set comprised literature values (excluding the constant) from Buiteveld (1990). These were shown in the previous chapter (see Table VII.4). The results of model calculation, using the estimated extinction coefficients from the present study, and the literature values of these coefficients, are shown in Figure (8.15). The simulated light extinction coefficients using the estimated specific extinction coefficients from the present study are generally lower than the estimated extinction coefficients using literature values from Buiteveld (1990). The difference is attributed to the lower specific extinction coefficients for TSS obtained in the present study. While the specific coefficients for algae and gilvin obtained in the present study are generally higher than literature values, they do not contribute much to the total extinction coefficient due to their low concentrations in Cape Bolinao. Using the average algal concentration and gilvin absorption coefficient, their total contribution is about 0.16 m^{-1}, roughly 40 % of the mean k_d (≈ 0.40 m^{-1}) in Cape Bolinao. On the other hand, the contribution of TSS (organic and inorganic) to k_d is, on the average, about 52 % (≈ 0.21 m^{-1}).

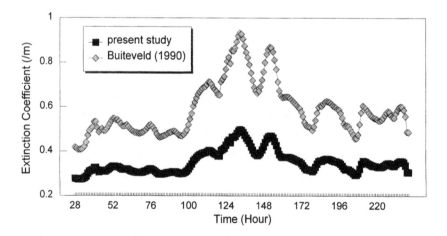

Figure 8.15. Simulated light extinction coefficients using the estimated specific extinction coefficients from the present study and literature values (excluding constant) from Buiteveld (1990).

From linear approximation, this would give a mean contribution of 0.03 m^{-1} for pure water which is within the mean background extinction as reported by Smith and Baker (1981).

8.4. Model Validation with Observed Data

Using observed data from Cape Bolinao, the numerical models developed in this study have been validated. While a comprehensive validation can not be achieved due to the lack of independent data which had not been used for calibration purposes, the results of the 'partial' validation made here can give insight into the usefulness and predictive capability of the numerical models. The results of this exercise are presented in the following sub-sections.

8.4.1. Current Velocities and Circulation Patterns

The numerical model for hydrodynamics has been run to simulate the two prominent seasons, i.e. the northeast (NE) and southwest (SW) monsoons. A uniform wind speed of 3 m s^{-1} (mean annual), representative for both seasons, was used. The mean wind direction used for the northeast monsoon is about 337.5° (as shown by meteorological observations), while for the southwest monsoon season, a mean wind direction of 225° is used. The model was run in conjunction with a tidal forcing which includes the four primary constituents, namely O_1, K_1, M_2 and S_2. The simulated depth-mean currents for both seasons and for both flood and ebb tides are presented in Figures (8.16-8.17). Also, the simulated (and observed) surface currents for both seasons are presented in Figures (8.18-8.19). Comparing the results, there is a reasonable agreement between the simulated and observed surface currents. The simulated and observed surface currents are of the same order of magnitudes ($\approx$ 10 cm s^{-1}). Regarding the current directions, the discrepancy (especially for the southwest monsoon season) can be attributed to the tide and variability in the wind direction. It should be noted that measurements of the surface currents could not be done simultaneously at all the sites. In this case, changes in the tidal phase and wind direction during measurements are manifested in the observed surface currents.

A further validation of the hydrodynamic model was made in terms of the simulated (tide) sea surface elevation (Figure 8.20). As shown, there is a close agreement between observation and model simulation. The minor differences in the sea surface elevation is due to the difficulty in describing the actual tide in the area. It should be noted that the model includes only the four primary tidal constituents. There are also natural oscillations of the sea surface with frequencies close to the frequencies of the involved tidal constituents. All these factors, contribute to the observed sea surface elevation.

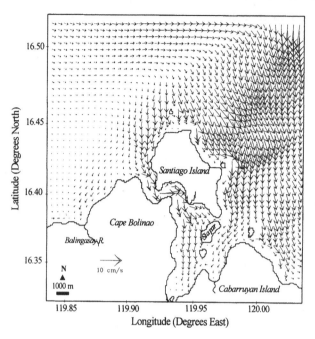

Figure 8.16a. Simulated circulation pattern (depth-mean currents) in Cape Bolinao during the northeast monsoon season (flood tide).

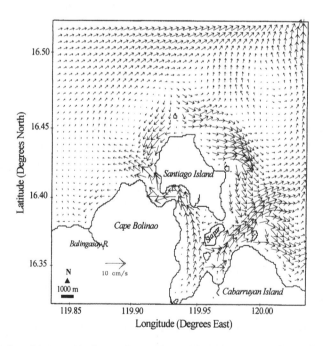

Figure 8.16b. Simulated circulation pattern (depth-mean currents) in Cape Bolinao during the northeast monsoon season (ebb tide).

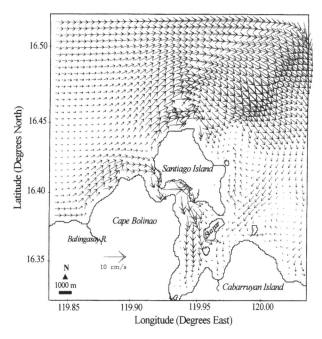

Figure 8.17a. Simulated circulation pattern (depth-mean currents) in Cape Bolinao during the southwest monsoon season (flood tide).

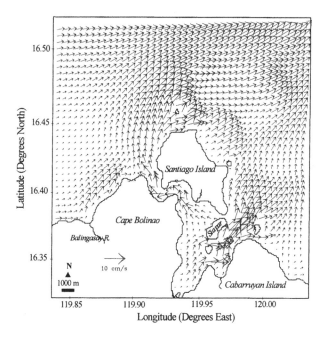

Figure 8.17b. Simulated circulation pattern (depth-mean currents) in Cape Bolinao during the southwest monsoon season (ebb tide).

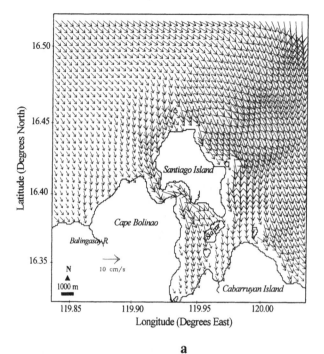

a

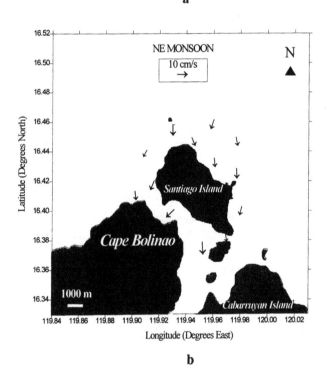

b

Figure 8.18. Predicted (a) and observed (b) surface currents during the northeast monsoon season. The predicted surface current components are calculated from $u_{surf}, v_{surf} = 1.5u, v + a_{x,y}/4$.

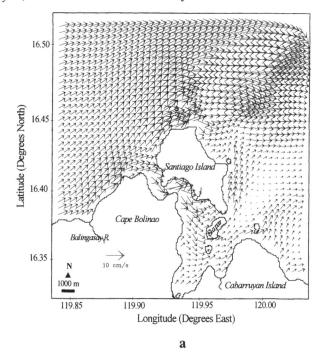

a

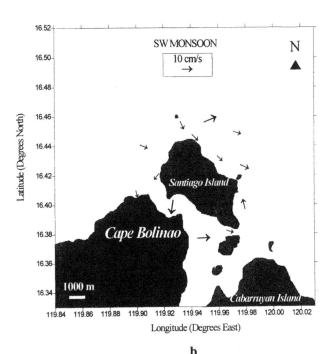

b

Figure 8.19. Predicted (a) and observed (b) surface currents during the southwest monsoon season. The predicted surface current components are calculated from $u_{surf}, v_{surf} = 1.5u, v + a_{x,y}/4$.

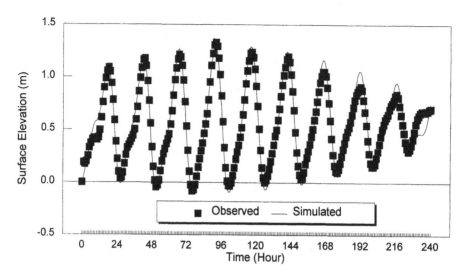

Figure 8.20. Observed and predicted tide (sea surface elevation) in Cape Bolinao.

8.4.2. Suspended Sediment Dynamics and Distribution Pattern

It was shown in the foregoing sensitivity analyses that the simulated TSS concentrations are sensitive to both waves and currents particularly during stormy conditions. To test the validity of the sediment transport model, the storm event (October 1994) in Cape Bolinao was simulated. During this period, surface winds of about 20 m s^{-1} were experienced in the area. The effects of both waves and currents as driving forces for resuspension and transport were included in the model run. The values of the different parameters which resulted from a series of sensitivity tests and calibration can be seen in Table (VIII.2).

Parameter	Description	'clay'	'silt'	'sand'
k_r (kg m^{-2} s^{-1})	resuspension constant	2.0×10^{-7}	3.0×10^{-6}	5.5×10^{-6}
τ_{cr} (N m^{-2})	critical shear stress	0.14	0.15	0.19
w_s (m s^{-1})	settling velocity	3.5×10^{-4}	5.5×10^{-4}	1.0×10^{-3}
α (m N$^{1/2}$)	resuspension parameter	0.5	0.5	0.5

Table VIII.2.   Parameter values used in the suspended sediment transport model.

The simulated depth-mean currents, significant wave heights and bottom shear stress during the maximum winds of the storm are shown in Figures (8.21-8.23). The simulated depth-mean currents during the storm went over 50 cm s^{-1} (Figure 8.21). The general circulation pattern shows an anticyclonic gyre northwest of Santiago Island. A strong southward flow in the reef can be seen. This circulation pattern is also shown by the previous model simulation using uniform winds from the NNW direction. The simulated significant wave heights are generally less than 1 m in the reef area (Figure 8.22). While higher wave-heights are expected in the deeper areas outside the reef, these simulated wave heights are deemed representative in most of the shallow reef areas of interest.

The (total) bottom shear stress went over 1 N m^{-2} during the maximum wind of the storm. Generally, the wave-induced stress is more dominant than the current-induced stress near the Bolinao coast (Figure 8.23). However, the current-induced stress can be comparable in magnitude to the wave-induced stress in some areas of the reef.

During the same period, the simulated sediment transport rates in Cape Bolinao are shown in Figure (8.24). The depth-averaged suspended load transport rates in Cape Bolinao are generally low ($<$ 500 g m^{-1} s^{-1}). This is attributed to the low TSS concentrations available for transport.

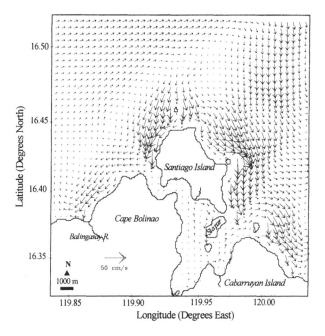

Figure 8.21. Simulated depth-mean currents (cm s^{-1}) in Cape Bolinao during the storm of October 1994. The resulting suspended sediment transport is partly due to this currents.

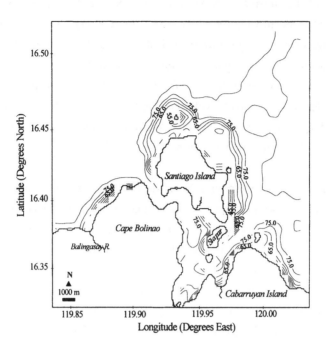

Figure 8.22. Simulated significant wave heights (cm) in Cape Bolinao during the storm of October 1994.

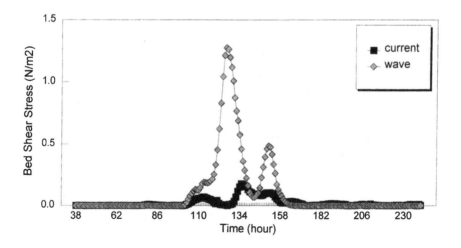

Figure 8.23. Simulated bottom shear stress (N m^{-2}) due to current and waves in Cape Bolinao during the storm of October 1994.

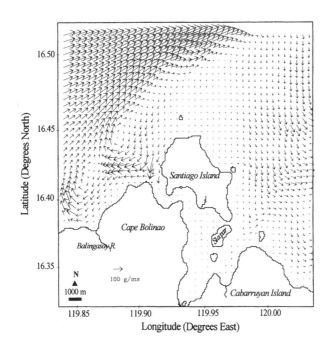

Figure 8.24. Simulated sediment transport rates (g s^{-1} m^{-1}) in Cape Bolinao during the storm in October 1994.

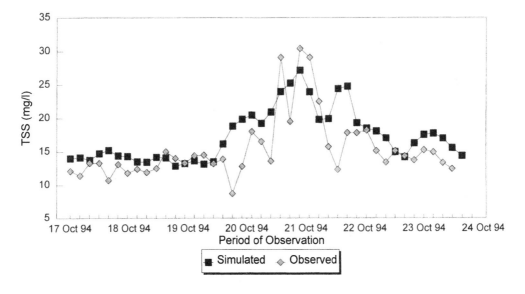

Figure 8.25. Simulated and observed TSS concentration (g m^{-3}) in Cape Bolinao during the storm of October 1994.

The unavailability of observed data on the significant wave heights, bottom shear stress, and suspended load transport during the storm precludes validation of these simulated results. However, field observations of TSS concentrations north of Cape Bolinao coast during this period (17-24 October 1994) are available and these are used for model validation (Figure 8.25). It can be seen that the model produces a reasonable agreement between the simulated and the observed TSS data. With the inherent uncertainties in the simulated significant wave characteristics, the associated bed shear stresses due to waves and currents, and the resuspension flux, the simulated magnitudes and variations in the TSS concentrations are not very different from the observed concentrations. If the observations are smoothed out, the simulated TSS concentrations approximate the observed concentrations.

8.4.3. Light Extinction

Using the estimated specific extinction coefficients for the combined reef-channel system, the light extinction coefficients corresponding to the northeast and southwest monsoon seasons were simulated in Cape Bolinao. Results of the simulations for both seasons can be seen in Figures (8.26). For an areal description of the light extinction coefficients, the results of the model run for both ebb and flood tide are shown in Figure (8.27-8.28). The simulated extinction coefficients were generally low (0.29 - 0.48 m^{-1}) but within the observed range of light extinction coefficients in Cape Bolinao. There is little difference (≈ 0.02 m^{-1}) in the magnitudes of k_d for both seasons, and for both flood and ebb tides. However, the

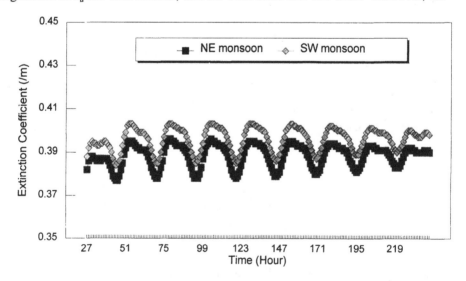

Figure 8.26. Time series of simulated light extinction coefficients in Cape Bolinao during the northeast and southwest monsoon seasons.

distribution pattern of the light extinction coefficients appears to change with the tide. This is also true for both seasons. The transport and redistribution of suspended fine sediments by tidal and wind-driven currents determine the simulated distribution of the extinction coefficients. Note the different circulation patterns during different seasons and tidal regimes (Section 8.4.1).

An actual validation of the light extinction coefficient model is made for the period 10-30 November 1994. Observed wind data in conjunction with a tidal forcing are used to force the hydrodynamic model. The sediment transport model simulates the three sediment fractions. In the light extinction model, the specific extinction coefficients obtained in the present study and literature values (excluding the constant) from Buiteveld (1990) were used. Here, distinction was made between the contributions of organic and inorganic fraction in the total suspended sediment concentrations. The algal concentration and gilvin absorption coefficient are set to their average values (in accordance with observed data). The result of the simulation is shown in Figure 8.29. As shown, there is a reasonable agreement between the simulated and observed light extinction coefficients using the set of specific extinction coefficients obtained from the present study. Using Buiteveld (1990) data resulted in very high light extinction coefficients. The discrepancy between the simulated and observed k_d could be due to uncertainty in the specific extinction coefficients.

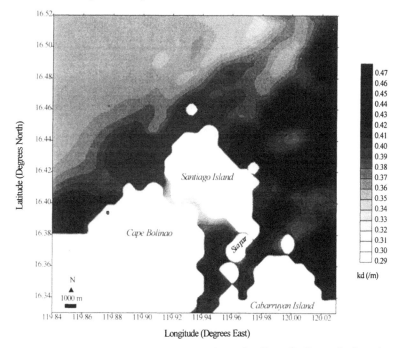

Figure 8.27a. Simulated light extinction coefficients in Cape Bolinao during the northeast monsoon season (flood tide).

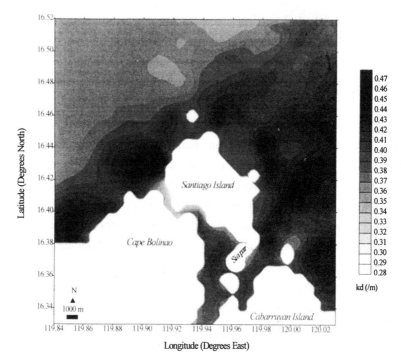

Figure 8.27b. Simulated light extinction coefficients in Cape Bolinao during the northeast monsoon season (ebb tide).

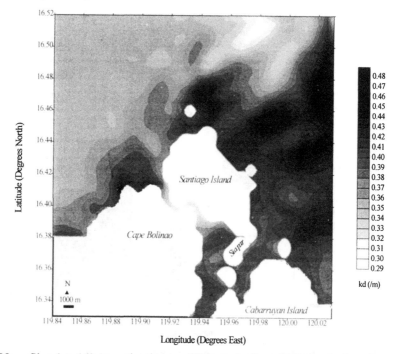

Figure 8.28a. Simulated light extinction coefficients in Cape Bolinao during the southwest monsoon season (flood tide).

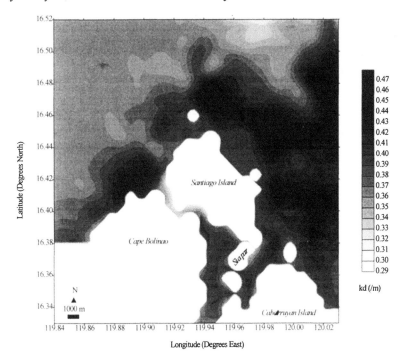

Figure 8.28b. Simulated light extinction coefficients in Cape Bolinao during the southwest monsoon season (ebb tide).

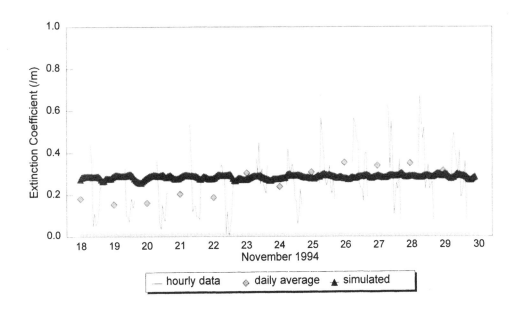

Figure 8.29. Simulated and observed light extinction coefficients in Cape Bolinao during the period 18-30 November 1994.

It can also be seen in Figure (8.29) that the observed extinction coefficients show a more dynamic character than the simulated extinction coefficients. Daily observations usually show higher extinction coefficients at sunrise and before sunset. This is attributed to the longer pathlength that the quanta of radiation traverse in the water column and the associated increase in the absorption and scattering coefficients when the sun is not at the zenith. The lack of dynamics in the modelled extinction coefficients is due to the fact that the sun is assumed to be perpendicularly above the water column (i.e. at the zenith).

Chapter 9

Simulations and Impact Assessment in Cape Bolinao and Lingayen Gulf

Using the numerical models developed in the present study, the impact of external sediment loads in Cape Bolinao and the Lingayen Gulf is investigated. The external sediment loads usually come from river discharges. Accelerated soil erosion during heavy precipitation, river bank erosion, tailings from mining activities, agricultural run-off, and domestic sewage all contribute to the bulk of the river loads. Partly, the deforestation in the watershed and land use changes for agriculture and urban expansion contribute to the high rate of erosion in the surrounding areas. As a consequence, the Lingayen Gulf is suffering from a high rate of sedimentation. As reported by Santos et al. (1989), the sedimentation rate in the gulf is estimated through nuclear isotope dating to be about 4.7 cm yr^{-1}

It is particularly important to investigate the transport of the fine suspended sediments coming from river sources. The transport and redistribution of these materials by wind and tide-driven currents can have a significant influence on the water quality of the gulf waters. For example, the underwater light extinction has been shown in the previous chapters to be strongly dependent on the concentrations and distribution patterns of the suspended material. Furthermore, there is the associated pollution of the gulf waters by nutrients and micropollutants adsorbed onto the suspended particulates. From an ecological viewpoint, the whole Lingayen Gulf ecosystem is endangered by siltation and eutrophication problems.

For the impact assessment undertaken in the present study, the transport and distribution pattern of the suspended material and the associated effect on the light conditions in Cape Bolinao and the Lingayen Gulf have been quantified by the numerical models described in the previous chapters. Two case studies are presented. The first one concerns the impact on the Bolinao reef of sediment discharges by rivers in the vicinity of Cape Bolinao itself.

Here, the transport of the suspended sediments (from rivers and resuspended bottom materials) and their impact on light extinction is investigated (Section 9.1). The second one concerns the impact of river discharges in the Lingayen Gulf. Here, the discharged river sediments are assumed to be conservative and thus function as a tracer showing the simulated transport patterns (Section 9.2). This case study is especially undertaken to determine whether river discharges in the south of the Lingayen Gulf can have an impact on the Bolinao reef system. The results of the numerical investigations are presented in the following sections.

9.1. Impact of External Sediment Loads in Cape Bolinao

The rivers near Cape Bolinao which are likely to have an impact on the Bolinao reef system include the Balingasay River, Guiguiwanen creek, Luciente River, Ensiong River in Santiago Island, and Sta. Rita River in Cabarruyan Island (see Figure 9.1). The contribution of these river systems to the external sediment loads are generally very low during the dry season (NE Monsoon). However, considerable TSS loads are usually discharged by these rivers during the rainy season (SW Monsoon). Field measurements in the Balingasay river during the start of the rainy season of 1995 showed a mean TSS concentration of 17.2 mg/l. This observation was undertaken during a very ordinary rain event, i.e. total precipitation was recorded at 22.2 mm in a rainy period of 4 hours. This gave an average precipitation rate of 5.6 mm hr^{-1}. With higher precipitation rates experienced in the area during the rainy season, river loads are expected to increase due to enhanced erosion rates. And with deforestation as an additional problem in the area, the associated higher extreme run-off rates also implies higher sediment loads.

9.1.1. Impact on the Levels of TSS Concentrations

In modelling the impact of river discharges, it was assumed that the fine TSS loads discharged by the rivers close to Bolinao are 50 mg/l each, except for the Balingasay River which is the biggest in the area, where a TSS concentration of 100 mg/l is assumed. The simulated TSS distribution pattern during the rainy season is shown in Figure (9.1). This is the simulation result after 6 months (SW monsoon season). It can be seen that the levels of TSS concentrations in the Bolinao reef are only slightly affected by the river discharges as the TSS levels (12-15 mg/l) are comparable to the result of the simulations without river discharges (see Chapter 8). The time series of simulated concentrations in Figure (9.2) shows this slight increase in the TSS levels. There is a gradual build up of TSS concentrations which is attributed to the river discharges. The impact of the river loads on the TSS levels in the reef at the end of the rainy season is generally less than 2 mg/l.

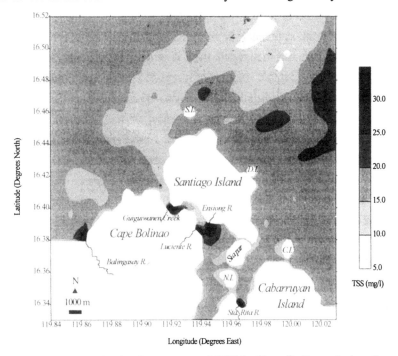

Figure 9.1. Simulated distribution pattern of TSS in Cape Bolinao during the rainy season. Combined effects of resuspension and sedimentation of bottom sediments and river discharges are included in the sediment transport model.

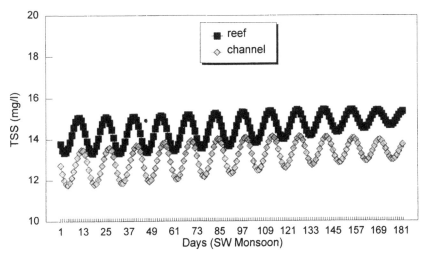

Figure 9.2. Simulated temporal variation of TSS concentration in the Bolinao reef during the rainy season. Combined effects of resuspension and sedimentation of bottom sediments and river discharges are included in the sediment transport model.

Assuming the external sediment load as a conservative tracer, an impact study of the river discharges was further undertaken in Cape Bolinao. This investigation is tantamount to assuming that only the rivers contribute to the suspended sediments around Cape Bolinao and that resuspension of bottom sediments is completely balanced by sedimentation. The sediment transport model was run to simulate the SW monsoon season (rainy) where rivers are expected to discharge considerable amounts of suspended sediments to the coastal system. The model was run for a 6-month duration (SW monsoon) and the final concentrations are shown in Figures (9.3-9.4). It can be seen that most of the 'external sediments' are transported along the Bolinao coast and into the channel area (Figure 9.3). The Bolinao reef (north of Santiago Island) receives a maximum TSS concentration of 2 mg/l. The time series of TSS concentrations in Figure (9.4) also shows the minimal influence of river discharges in the reef area. In most sites, a TSS contribution of less than 2 mg/l can be seen. This minimal influence of river discharges in the reef can be attributed to the influence of the tide in the area. It was shown in the hydrodynamical simulations (Chapter 8), that the tidal circulation around Cape Bolinao favors the advection of suspended matter east or west of the Santiago Island, bypassing most of the reef area north of the Santiago Island.

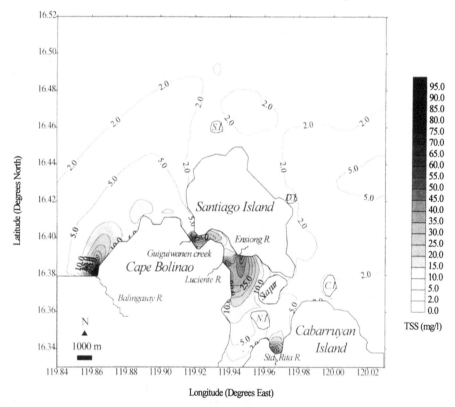

Figure 9.3. Simulated distribution pattern of a conservative tracer in Cape Bolinao after the entire SW monsoon season. The contours represent approximate TSS concentrations in mg/l.

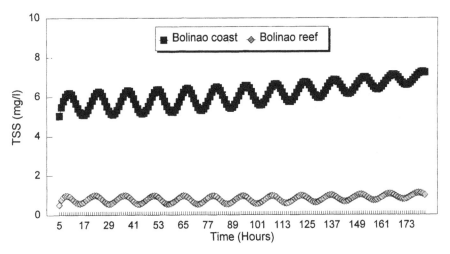

Figure 9.4. Time series of tracer concentrations in the coast and the reef areas of Cape Bolinao. The simulated concentrations are in mg/l.

9.1.2. Impact on Light Extinction

Using the predicted TSS concentrations in Figure (9.1), the light extinction in Cape Bolinao is simulated. Here, a constant algal contribution is assumed using an average chlorophyll-a concentration of 0.45 $\mu g/l$. Also, an average gilvin absorption coefficient of 0.95 m^{-1} and a background extinction of 0.1 m^{-1} for pure water are assumed. Although short term (weekly to monthly) fluctuations of algal and gilvin contributions have been observed, the use of constant partial extinction coefficients for these components allows the assessment of the impact of external sediment inputs on the light extinction in Cape Bolinao. Assuming that the sediment inputs have the same specific extinction coefficients as the internal sediments, the simulated light extinction coefficients are shown in Figures (9.5-9.6). The extinction coefficients show very little influence of the river discharges in the reef as the levels of k_d (0.35 - 0.5 m^{-1}) shown are close to those obtained in simulations without river discharges (see Chapter 8). With river discharges, there is only a very slight increase (much lower than 0.1 m^{-1}) in the k_d values in the Bolinao reef.

Simulations of the impact of river discharges (assumed as a conservative tracer) on the light conditions in Cape Bolinao also showed little influence of external sediment loads on the light extinction in the reef (Figures 9.7-9.8). Because of the expected transport pattern of suspended matter in the area around Cape Bolinao, higher extinction coefficients can be seen in the Bolinao coast and in the channel than in the reef area. This indicates that even when river discharges are minimized, light conditions in the reef will remain almost unchanged.

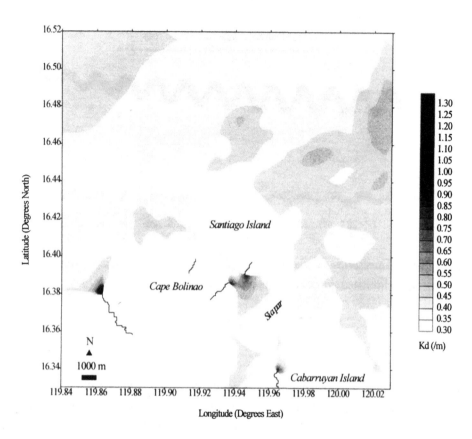

Figure 9.5. Simulated light extinction coefficient in Cape Bolinao during the rainy season. River discharges and resuspension and sedimentation of bottom sediments are included.

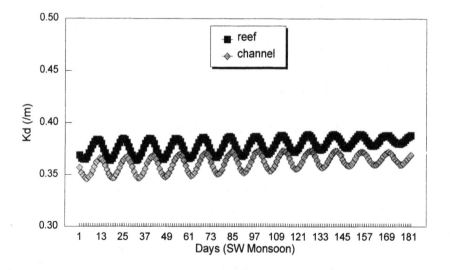

Figure 9.6. Time series of simulated light extinction coefficient in the Bolinao reef during the rainy season. River discharges have been included in the simulation.

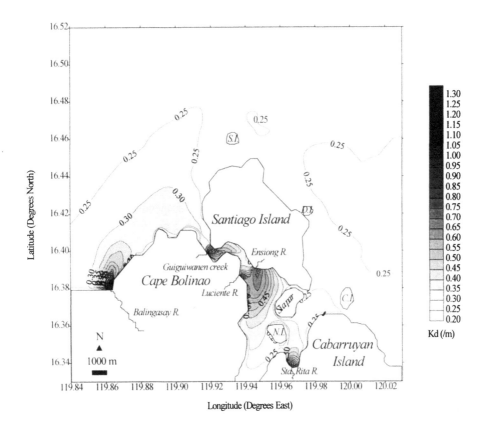

Figure 9.7. Simulated light extinction coefficients in Cape Bolinao due to non-settleable river inputs during the entire SW monsoon season. The contours represent approximate extinction coefficients in m⁻¹.

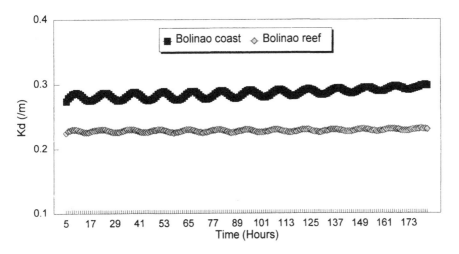

Figure 9.8. Temporal variation of the simulated light extinction coefficients in Cape Bolinao due to a conservative tracer.

9.2. Impact Assessment in the Lingayen Gulf

One of the questions posed during the early stages of this study is whether the sediment discharges by the river systems in the Lingayen Gulf have an impact on the Bolinao reef system. To answer this question, an integrated simulation of the hydrodynamics and sediment transport in the gulf has been carried out. The results of the modelling study on the hydrodynamics, and implications for the transport of suspended matter are presented in Section (9.2.1). Using the predicted flow velocities to force the suspended sediment transport model, the transport of river loads (assumed as a conservative, non-settleable tracer) and the probable (maximum) impact on the levels of suspended matter in the Lingayen Gulf (including Cape Bolinao) was investigated (Section 9.2.2).

9.2.1. The Hydrodynamics of the Lingayen Gulf

The coarse resolution model described in Chapter 4 is used to simulate the long-wave currents in the Lingayen Gulf. The individual contribution as well as the interaction of the wind and tide-driven currents in the general circulation of the gulf are considered in the present study. While a uniform wind is used to assess the wind-driven currents in the gulf, a realistic tidal forcing, obtained from observations in Cape Bolinao, is used to simulate the tide-driven currents. Similar to the fine resolution model for Cape Bolinao, the coarse resolution model applied to the gulf used a constant diffusion coefficient ($A_h = 10$ m^2 s^{-1}) in the momentum equations.

9.2.1.1. Wind-Driven Circulation

Using uniform winds for the two prominent seasons (SW and NE), the wind-driven circulation patterns in the Lingayen Gulf are shown in Figures (9.9-9.10). A uniform wind speed of 3 m s^{-1} for both seasons has been used. The wind directions assumed for the NE and SW monsoon seasons are 337.5° and 225° respectively. The tidal forcing was further set to zero.

During the SW monsoon season, the wind-driven circulation pattern shows a general eastward flow along the southern coast of the Lingayen Gulf (Figure 9.9). At the eastern coastal areas, a northward flow is observed. This is compensated by a southward flow in the center of the gulf. In the western part of the gulf, an anticyclonic gyre centered about 9 km east of Cabarruyan Island develops.

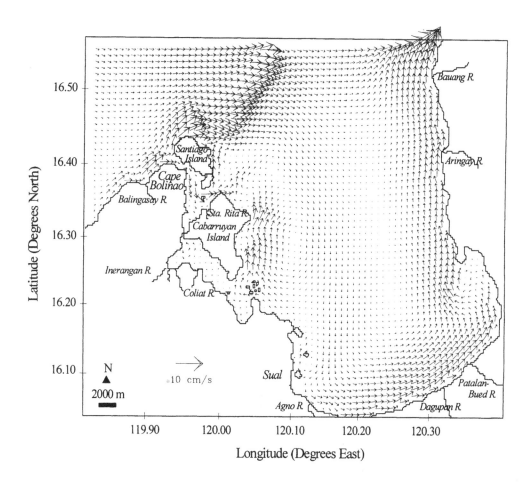

Figure 9.9. Simulated wind-driven circulation pattern in the Lingayen Gulf during the southwest monsoon season.

During the NE monsoon season, two pronounced circulations can be seen in the Lingayen Gulf namely, an anti-cyclonic gyre in the western part of the gulf and a cyclonic gyre in the eastern part (Figure 9.10). Also, there appears a weak anticyclonic gyre northwest of Cape Bolinao during this season. This small gyre (also predicted by the fine resolution model) vanishes during the southwest monsoon season.

In general, the simulated depth-mean current magnitudes for both seasons are less than 10 cm s^{-1}. If the surface currents are calculated according to Equation (8.3), it can be shown that these simulated results are also in good agreement with observations in Cape Bolinao (see Chapter 7). A further validation for the Lingayen Gulf circulation is difficult because most measurements were available only for Cape Bolinao which is the main area of interest.

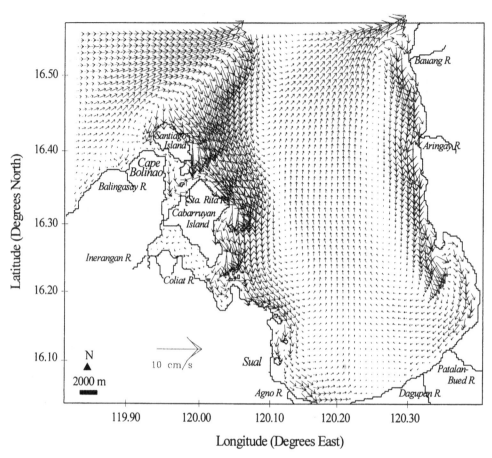

Figure 9.10. Simulated wind-driven circulation pattern in the Lingayen Gulf during the northeast monsoon season.

9.2.1.2. Tide-Driven Circulation

The simulated tide-driven circulation patterns in the Lingayen Gulf are shown in Figures (9.11-9.12). The results show a general southward mass transport during flooding with the tidal currents having magnitudes of less than 10 cm s^{-1} in most parts of the gulf (Figure 9.11). Similar to the prediction of the fine resolution model (see Chapter 8), the flood currents appear to penetrate inside the channel bordering Cape Bolinao and Santiago Island. The flood currents enter the channel on both sides of the Santiago Island. In the Tambac Bay west of Lingayen Gulf, strong flood currents can be seen. The channel area between the southern tip of the Cabarruyan Island and the mainland appear to be a tidal inlet where flood currents attain maximum speeds of over 30 cm s^{-1}. The current pattern is reversed during ebbing (Figure 9.12). A general northward transport in the Lingayen Gulf can be seen. Also similar to the prediction of the fine resolution model, the ebb currents at the channel

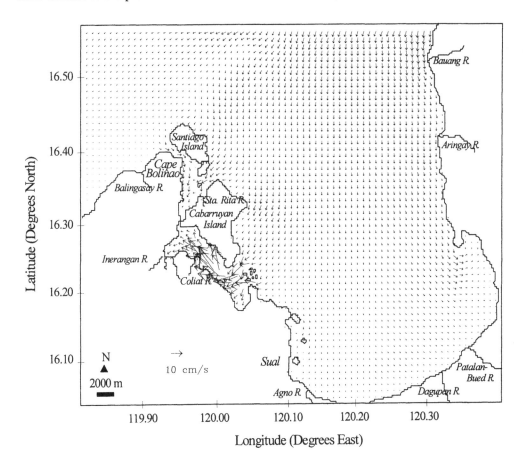

Figure 9.11. Simulated tide-driven circulation pattern in the Lingayen Gulf at flood tide.

of Cape Bolinao show a general northward flow exiting on either side of the Santiago Island. In the Tambac Bay, a strong outward flow towards the Lingayen Gulf can be seen. The relatively higher flow velocities in this area as compared to the rest of the gulf waters are attributed to the difference in the sea surface elevation inside Tambac Bay and the Lingayen Gulf. The simulated ebb currents are slightly stronger than the flood currents (see scale). Except in the Tambac Bay, the flood currents in the Lingayen Gulf are generally much less than 10 cm s^{-1}. The ebb currents approaches 10 cm s^{-1} in some areas of the gulf such as the Bolinao channel.

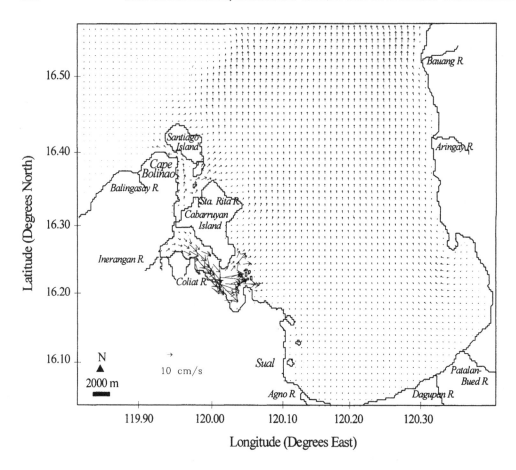

Figure 9.12. Simulated tide-driven circulation pattern in the Lingayen Gulf at ebb tide.

9.2.1.3. Wind and Tide-Driven Circulation

The interaction between the wind and the tide has been simulated with both driving forces included in the hydrodynamic model. It should be noted that the combined wind and tide-driven currents (and not just the independent action of the wind and the tide) are the true and realistic conditions when describing the suspended sediment transport in the Lingayen Gulf. The results of the simulation, assuming a southwesterly wind direction, are shown in Figures (9.13-9.14) for flood and ebb conditions respectively.

The general circulation during flood tide (Figure 9.13) shows a strong curvature of the currents past Cape Bolinao from a northeastern to a southeastern direction. Furthermore, an amplification of the current velocities in the area northeast of Cape Bolinao can be seen. It should be noted that this is a shallow area (called a barrier reef by McManus et al. 1994)

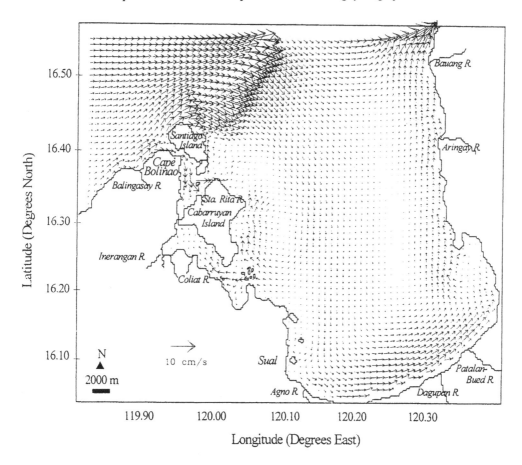

Figure 9.13. Simulated circulation pattern in the Lingayen Gulf with a uniform SW wind of 3 m s⁻¹ at flood tide.

and hence surface stress effects become significant. Also, due to the sudden decrease in the depth distribution, the associated gradient in the sea surface elevation contributes to an increase in the flow velocity.

In the middle of the gulf, a general southward flow can be seen (Figure 9.13). This is counteracted by a northward flow along the eastern coasts. A series of anticyclonic gyres forming a vortex street can be seen in the western coastal areas. The most pronounced of these gyres can be seen east of Cabarruyan Island. Off Santiago Island, two weak anticyclonic circulations can also be seen. These are attributed to the non-linear interaction of the wind and the tide. It should be noted that these circulation patterns do not appear in the separate simulations for wind and tide-driven flows.

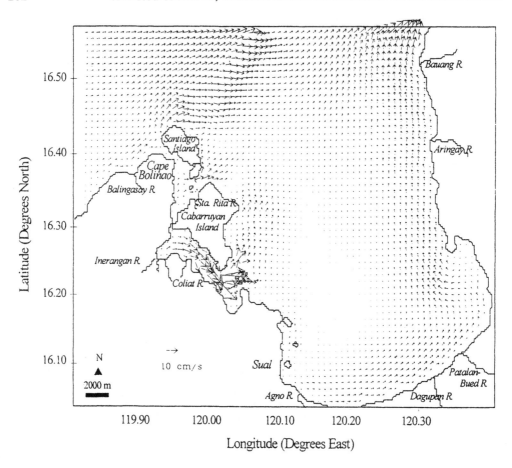

Figure 9.14. Simulated circulation pattern in the Lingayen Gulf with a uniform SW wind of 3 m s^{-1} at ebb tide.

At ebb tide, the vortices along the west coast of the Lingayen Gulf vanish and the circulation pattern in most areas shows a general mass transport in the northeast direction (Figure 9.14). Furthermore, similar to Figure (9.13), there is a general eastward transport in the south of the gulf. It should be noted that this has an important consequence on the transport of sediments discharged by the big rivers in this area, i.e. the western part of the gulf is barely affected by the river discharges in the south. The magnitudes of the current is higher at ebb tide. Especially in the Tambac Bay, the wind and tide-driven currents attain higher values, often exceeding 40 cm s^{-1}. The amplification of the current velocity in this area is due to the complementary action of the wind and the tide. Note that at ebb tide, strong currents exit Tambac Bay. The wind, which comes from the southwest in this case, amplifies the ebb currents in the bay. The interaction between the wind and the tide is further shown in the area northeast of Cape Bolinao. Here, the generally northward tidal flow is modified to an eastward flow due to the prevailing wind direction.

9.2.2. Transport of a Conservative Tracer in the Lingayen Gulf

There are five major rivers discharging sediments into the Lingayen Gulf namely, Agno, Dagupan, Patalan-Bued, Aringay and Bauang Rivers. There are also some other small rivers located in the western coast of the gulf with a potential contributions to the external sediment loads in the area. These include Inerangan, Garita, Barcadero and Coliat rivers which all drain in Tambac Bay west of the Lingayen Gulf. The Agno and the Patalan-Bued systems are considered to be major contributors to the siltation problem in the gulf. All other rivers contributing external sediment loads, derive these from either deforestation actions, mining or domestic sewage.

To simulate more or less realistic sediment transport patterns in the Lingayen Gulf, the sediment concentrations and flow discharges by the rivers are needed as input. The mean flow discharges during the rainy season, including the drainage areas of some rivers, are listed in Table IX.1.

River	Drainage Area (km^2)	Discharge (10^6 m^3 d^{-1})
Agno[a]	5952.0	6664.0
Dagupan[a]	897.0	1002.0
Patalan-Bued[a]	347.0	388.0
Alaminos[a,1]	200.0	224.0
Balingasay[b]	72.0	6.0

[a] source: NWRC Philippines, 1976.

[b] source: EIA, 1994.

[1] comprises Inerangan, Garita, Barcadero and Coliat rivers.

Table IX.1. Major rivers contributing to the sediment loads in Lingayen Gulf and Cape Bolinao.

For TSS concentrations, an average of about 250 mg/l was measured near the mouth of the Agno river (EIA 1994, AGNO River Flood Control). For the rivers in the west coast

(Alaminos), a TSS concentration close to 100 mg/l was measured near the river mouth during the dry season. The present levels of suspended sediments in the other rivers are not available due to the absence of field measurements. However, reasonable assumptions can be made. It can be assumed that the levels of TSS concentrations are increased due to increased river loads from increased mining activities east of the gulf and increased rates of erosion due to urban expansion and agricultural activities. Assuming that both the Agno and Patalan-Bued river systems contribute TSS concentrations of about 500 mg/l each (of the fine fractions), and assuming further that all the other small river systems contribute 100 mg/l to the marine waters of the Lingayen Gulf, the numerical model was run to simulate and predict the transport and distribution pattern of these suspended sediments. The sediment transport model assumes that the suspended sediments are conservative, i.e. decay and production of solids are neglected. So, a balance between sedimentation and resuspension of these suspended sediments is assumed. The absence of a bottom sediment model makes this a gross simplification but nevertheless also a reasonable assumption since settling sediments are usually resuspended again by tidal and/or wind-driven currents and become available for transport. Furthermore, it is the slowly settling, fine particulates which are significant when considering light extinction (see Chapter 6).

The simulated distribution patterns of a conservative tracer (representing the river sediments) for a 1-year duration are shown in Figures (9.15-9.17). The hydrodynamic model was driven using the annual mean wind speed and direction in conjunction with the tide. It can be seen that during the northeast monsoon (NE) season (Figures 9.15), Cape Bolinao is hardly affected by the river discharges in the south and the east of the Lingayen Gulf. The sediment loads coming from the big rivers in the south are generally transported along the central and eastern parts of the gulf. It is the rivers in the vicinity of Cape Bolinao (i.e. Balingasay and Sta. Rita) that contribute more to the suspended sediment concentrations in the cape, albeit minimal due to the general circulation patterns in the area around the cape. This effect has been shown in the previous simulations in Cape Bolinao (Section 9.1).

The simulation result at the peak of the southwest monsoon season (after 9 months) shows that the bulk of the river sediments from the southern part of the gulf would be transported towards the eastern coasts (Figure 9.16). This is also attributed to the general circulation patterns during this season. It is further shown that Cape Bolinao is more vulnerable during this season. The (already) suspended materials could be transported in the surroundings of the cape due to the prevailing wind and tide-driven currents around the area. However, the levels of TSS concentrations in the Bolinao reef are not significantly increased even during this period (rainy season). This is partly attributed to the dispersion of suspended matter that can occur at the open sea boundaries. Furthermore, the effect of the tide (flushing) in Cape

Bolinao can significantly reduce the concentrations of suspended matter in the open reef system. Suspended sediments appear advected (transported) out of the cape into the open sea.

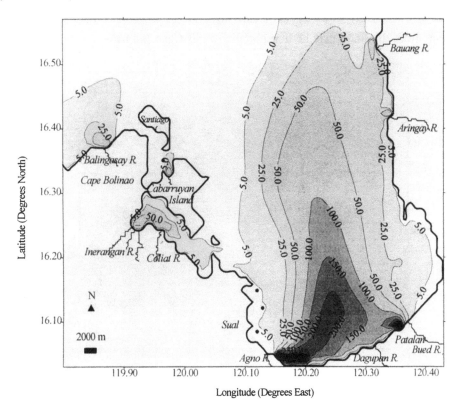

Figure 9.15. Simulated distribution pattern of a conservative tracer in the Lingayen Gulf after 3 months (NE monsoon season). The contours represent approximate TSS concentrations in mg/l.

The result for a 1-year simulation (also representing the NE monsoon season) is further shown in Figure (9.17). Similar to (Figure 9.15), it can be seen that the bulk of the suspended matter would be transported along the axis of the Lingayen Gulf from the south to the north. The influence of the big rivers in the southern parts of the gulf to Cape Bolinao also appears minimal.

The results of the simulation could not be verified due to the absence of a field experiment on tracer distribution patterns. However, with the validated circulation and transport models developed in the present study, the general distribution patterns are most likely as shown. The probability remains, however, that the sediment loads by the big rivers in the south of the Lingayen Gulf incidentally can have a significant impact on the Bolinao reef system.

This can be true especially during the SW monsoon season when storm winds from varying directions can influence the dispersion patterns of suspended matter. For example, if a storm crosses south of the Lingayen Gulf and remains quasi-stationary in the southwest, southeasterly to easterly winds can drive currents, and as a probable consequence, suspended sediments from the south may be transported towards Cape Bolinao.

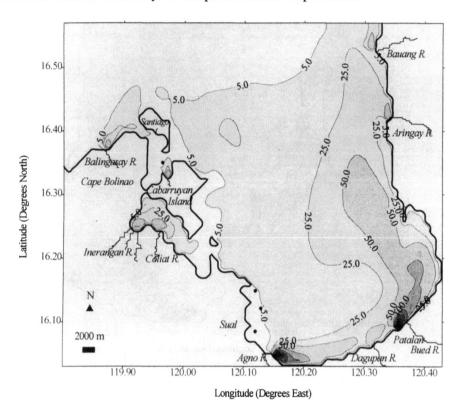

Figure 9.16. Simulated distribution pattern of a conservative tracer in the Lingayen Gulf after 9 months (SW monsoon season). The contours represent approximate TSS concentrations in mg/l.

The simulated distribution patterns of suspended matter have an important implications on the underwater biota in the entire Lingayen Gulf. The reduced light penetration due to the suspended matter can significantly affect the seagrass population in the area. The seagrasses in silted areas are becomingly scarce and this is partly due to relatively high light extinction coefficients in those areas. The impact of transported suspended matter in the Lingayen Gulf is not only confined to increased light extinction but also to eutrophication and heavy metal pollution of the affected marine waters. Nutrients and heavy metals tend to flocculate with the fine sediment fractions. Through suspended sediment transport by wind and tidal currents, transportation and redistribution of micropollutants in the gulf waters occur. The impact of these water quality problems is not only confined to the decline of seagrasses but also to the decline of the fishery resources in the whole Lingayen Gulf.

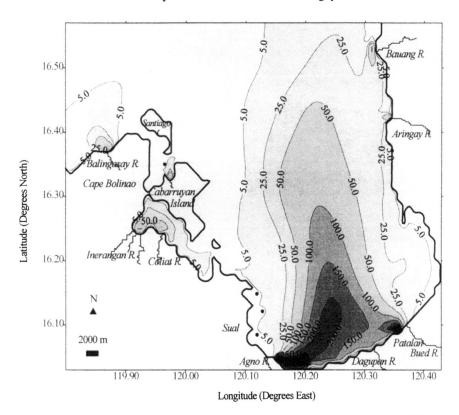

Figure 9.17. Simulated distribution pattern of a conservative tracer in the Lingayen Gulf after 1-year (NE monsoon season). The contours represent approximate TSS concentrations in mg/l.

Chapter 10

Summary and Conclusions

Observational and modelling studies on the hydrodynamics, sediment transport process and light extinction dynamics have been carried out in Cape Bolinao. The cape is situated at the mouth of the Lingayen Gulf at the northwestern coast of the Philippines. Typical of fringing coral reefs, it is abundant with various flora and fauna species such as seagrasses, seaweeds and corals. Siltation and eutrophication are threatening the biodiversity of this marine ecosystem. For this reason, an intensive study of relevant physical environmental factors has been carried out with the following objectives:

1. To provide a quantitative description of the general circulation patterns, sediment transport processes and light extinction in the marine waters off Cape Bolinao.

2. To develop an integrated numerical model of the hydrodynamics, sediment transport and light extinction at Cape Bolinao and adjacent waters for environmental impact studies.

The observational part of this study, which consisted of field measurements and laboratory experiments, was undertaken in a period of nearly two years from August 1993 to June 1995. Several measurement sites around Cape Bolinao were selected representative for most of the reef areas abundant with seagrasses. These seagrasses were the subject of a separate ecological study for which the light conditions were supposed to be an important input to be provided by the present study. Different frequencies of measurements, related to the time scales of the relevant processes and variables of interest, were employed in the observational program. These include a high-frequency hourly measurement, weekly measurement and quarterly measurement. For the high-frequency measurements, continuous observations of current speed and direction, total suspended solids (TSS) concentration, downward irradiance, and water level variation were executed at four designated sites in the reef using a platform. Except for TSS concentration, which was measured in intervals of 4 hours due to instrumental limitations, measurement of the other variables was undertaken in intervals of 5 minutes and were averaged hourly. Once in every 1 ½ months, the platform equipped with the automatic instruments was transferred to one of the 4 sites. This made it possible to cover both the dry (NE monsoon) and the rainy (SW monsoon) seasons in all the designated sites. For the weekly measurement, measurement of surface currents, temperature, salinity, total suspended solids concentrations, ash-free dry weight concentrations (of the suspended sediments), sedimentation flux, gilvin absorption coefficient, phytoplankton concentration (measured as chlorophyll-a) and downward irradiance were executed at 13 sites around Cape Bolinao. This provided data for analysis of the monthly and seasonal variability of relevant variables. For the quarterly measurements, sampling and analyses of bottom sediments were undertaken. Grain size distribution and organic content of the sediment samples were analyzed. For both the weekly and quarterly measurements, fractionation experiments to determine the size distribution and light attenuation characteristics of a mixture of sediments (bottom and suspended) were executed. Additional measurements on relevant meteorological variables such as wind (speed and direction) and rainfall were also undertaken.

The results of the field measurements on the hydrodynamics of Cape Bolinao showed that the currents are governed by the non-linear interaction of the wind and the tide. During normal conditions of no storms, surface currents are generally low in the reef ($\approx$ 10 cm s^{-1}). Tidal currents are of similar magnitudes except in some tidal inlets and the channel between Santiago Island and the mainland, where flow velocities often exceed 10 cm s^{-1}. Measurement of the water level variation showed a mixed tide with a dominant diurnal characteristic near Cape Bolinao. Once a fortnight, the semi-diurnal component in the tidal excursion became evident. The tidal range was about 1 m but was often exceeded during spring tide. The salinity of the reef waters had a distinct seasonal variation. Due to freshwater input, minimum salinity values (S < 30) were observed in the reef during the rainy season (SW monsoon). On the other hand, maximum salinity values (S > 33) were recorded during the dry season (NE monsoon) due to enhanced evaporation and negligible freshwater input. The temperature in the shallow reef waters showed a strong diurnal and seasonal fluctuation. This was attributed to the variation in the amount of insolation received. Concerning the suspended sediments, low TSS concentrations ($\approx$ 15 g m^{-3}) were observed around Cape Bolinao. Variations in the TSS concentrations during normal conditions of low wind speed was primarily dictated by tidal currents. During stormy conditions, the effect of the wind on the variability of the TSS concentration became significant. The sedimentation flux in the channel had a mean value of 3.2 g m^{-2} hr^{-1}. Lower sedimentation flux, with a mean of 0.7 g m^{-2} hr^{-1}, was observed in the reef area. Observations of the phytoplankton concentrations in the Bolinao reef showed very low chlorophyll-a content ($\approx$ 0.45 mg m^{-3}). However, higher algal concentrations, often exceeding 1 mg m^{-3}, were observed in the channel areas. The gilvin absorption coefficient (at 380 nm) showed a low average value of 0.95 m^{-1}. The organic matter content of the suspended sediments showed a mean value of 30 % during the whole period of observation. The bottom sediments around Cape Bolinao were heterogeneous. The reef areas showed more coarse than fine sediments while the opposite was observed in the channel areas where fine sediments were more dominant. The organic matter content of the bottom sediments were generally lower in the reef areas ($<$ 5 %) than in the channel areas (mostly 5 - 20 %). The light extinction coefficients in Cape Bolinao were low. A mean k_d of 0.40 m^{-1} was obtained during the whole period of observation. This low extinction coefficient was attributed to the low concentrations of suspended sediment, algae and gilvin in the area.

The modelling part of this study focused on the development of a set of prognostic and diagnostic finite-difference numerical models of the hydrodynamics, suspended sediment transport and light extinction dynamics. Two independent models were developed namely, a fine-resolution model of the main area of interest (around Cape Bolinao) and a coarse-resolution model of the whole Lingayen Gulf. The fine-resolution model was developed to describe and predict local processes while the coarse-resolution model was developed to understand transport processes in the gulf. The hydrodynamic model applied to both areas is a quasi three-dimensional model in the sense that current velocities at any depth can be estimated from model predictions. Realistic driving forces from the wind and the tide were used in this model. The tidal forcing consists of the four primary tidal constituents responsible for the water level variation in the area. These include the O_1, K_1, M_2 and S_2 tides. Using this prognostic hydrodynamic model for wind and tide-driven currents in conjunction with a diagnostic model for surface waves, a suspended sediment transport model was developed. The suspended sediment transport model is third-order accurate in space and time. With the addition of resuspension and sedimentation as source and sink terms, this model described the advection and diffusion of suspended sediments in the Cape Bolinao and the Lingayen Gulf. Like the hydrodynamic model, the sediment transport model was also based on a finite difference scheme in its solution approach. This third-order prognostic model can handle sharp gradients of sediment concentrations and does not suffer much numerical oscillations as long as the hydrodynamic forcing is numerically stable. Furthermore, there is very little mass falsification (e.g. mass is conserved) from this type of model which makes it desirable to use when investigating transport processes. The predicted sediment concentrations from this model, together with known contributions of algae, gilvin and pure water, were used in a diagnostic model for light extinction. This model is based on the assumption that the total light extinction coefficient is a linear function of the contributions of the different optically active components. Averaged algal concentration and gilvin absorption coefficient, obtained from field and laboratory measurements, were used in the simulations. Using linear approximation, it was shown that algae and gilvin contribute comparably ($\approx$ 20% each) to the extinction of light in Cape Bolinao. Inorganic and organic TSS (less algae) contribute 52 %, while pure water contributes about 8 % to the total light extinction coefficient.

A series of sensitivity analysis and calibration was performed with the numerical models. With the parameter values established through manual calibration, a reasonable agreement between observations and model simulations was achieved. This was especially true for current velocities, sea surface elevation, total suspended sediment concentrations and light extinction coefficients where direct observations were compared with model predictions.

Using the integrated model, the impact assessment studies done for Cape Bolinao showed little influence of the rivers on the levels of TSS concentrations and light extinction in the Bolinao reef. Sediment discharges in the vicinity of the cape appeared to be advected through the channel and away from the reef area. Furthermore, river discharges in the south of the Lingayen Gulf also showed minimal influence on the Bolinao reef system. From the model simulations, it was shown that more than 90% of the TSS concentration in the·reef was contributed by internal and local fluxes of resuspension and sedimentation of bottom sediments while less than 10 % was contributed by external (river) inputs. The slight effect of river loads was mainly attributed to the general circulation patterns in the Lingayen Gulf and the Cape Bolinao marine environmental systems.

References

Aalderink, R.H. L. Lijklema, J. Breukelman, W. van Raaphorst and A.G. Brinkman, 1984. Quantification of wind-induced resuspension in a shallow lake. Wat. Sci. Tech. Vol. 17. pp. 903-914.

Abbott, M. B., J. Larsen and J. Tao, 1985. Modelling circulations in depth-integrated flows Part 1: The accumulation of the evidence. J. Hydr. Res. Vol. 23, No. 4. p. 309 - 326.

Ahsan, A.K.M.Q., M.S. Bruno, L.Y. Oey and R.I. Hires, 1994. Wind-driven dispersion in New Jersey. J. Hydr. Eng. Vol. 120, No. 11. 1264 - 1273.

Andrews, J.C. and G.L. Pickard, 1990. The physical oceanography of coral-reef systems. In: Z. Dubinsky (ed), Coral reefs, ecosystems of the world. Elsevier, Amsterdam. 550 p.

Andrews, J.C., S. Gay and P.W. Sammarco, 1988. Influence of circulation on self-seeding patterns at Helix Reef - Great Barrier Reef. Proc. 6th Intl. Coral Reef Symp., Australia. Vol 2., 469-474.

Bartholdy, J. and P. P. Madsen, 1985. Accumulation of fine-grained material in a Danish tidal area. Marine Geology. Vol. 67. p. 121 - 137.

Baker, E. T. and J. W. Lavelle, 1984. The effect of particle size on the light attenuation coefficient of natural suspensions. J. Geophys. Res. Vol. 89, No. C5, p. 8197 - 8203.

Balotro, R. S., 1992. Development of a barotropic numerical model for the Lingayen Gulf. MSc. Thesis. Department of Meteorology and Oceanography. University of the Philippines, Diliman, Quezon City.

Blom, G. E.H.S. Van Duin, R.H. Aalderink, L. Lijklema and C. Toet, 1992. Modelling sediment transport in shallow lakes - interactions between sediment transport and sediment composition. Hydrobiologia 235/236 pp 153-166.

Blom, G. and C. Toet, 1993. Modelling sediment transport and sediment quality in a shallow Dutch lake (Lake Ketel). Wat. Sci. Tech. Vol. 28, No. 8-9, pp. 79-90.

Blom, G., E.H.S. van Duin and L. Lijklema, 1994. Sediment resuspension and light conditions in some shallow Dutch lakes. Wat. Sci. Tech. vol. 30, No. 10, pp 243-252.

Blom, G., E. H. S. van Duin and J. E. Vermaat, 1994. Factors contributing to light attenuation in Lake Veluwe. In: W. van Vierssen et al. (eds). Lake Veluwe, a macrophyte-dominated system under eutrophication stress. pp 158-174. Kluwer Academic Publ., Netherlands.

Blom, G. and D. L. Anderson, 1996. Modelling sediment transport in the EAA: Theory, Model Structure, and Preliminary Results. Paper presented at the SFWMD. Everglades, Florida.

Blumberg, A. F., R. P. Signell and H. L. Jenter, 1993. Modelling transport processes in the coastal ocean. J. Marine Env. Eng. Vol. 1. p. 31 - 52.

Bode, L. and R. J. Sobrey, 1985. Initial transients in long wave computations. J. Hydr. Eng. Vol. 110, No. 10. p. 1371 - 1397.

Bouws, R., 1986. Verwachting van zeegang door middel van groeicurves: bevindingen verkregen aan de hand van de Markermeerdataset. KNMI: 00-86-33. De Bilt, The Netherlands.

Bricaud, A. and D. Stramski, 1990. Spectral absorption coefficients of living phytoplankton and nonalgal biogenous matter: A comparison between the Peru upwelling area and the Sargasso Sea. Limnol. Oceanogr. Vol. 35, No. 3, p. 562 - 582.

Briones, N. D., 1987. Mining Pollution: the case of the Baguio Mining District, the Philippines. Environmental Management, Vol 11, No. 3. pp. 335 - 344.

Buiteveld, 1990. Uitzicht, model voor berekening van doorzicht en extinctie. Rijkswaterstaat, Dienst Binnenwateren/RIZA, nota nr. 90.058. Lelystad.

Campbell, J. W. and T. Aarup, 1989. Photosynthetically available radiation at high latitudes. Limnol. Oceanogr. Vol. 34, No. 8. p. 1490 - 1499.

Carder, K. L., R. G. Steward, G. R. Harvey and P. B. Ortner, 1989. Marine humic and fulvic acids: Their effects on remote sensing of ocean chlorophyll. Limnol. Oceanogr. Vol. 34, No. 1. p. 68 - 81.

Celik, I. and W. Rodi, 1988. Modeling suspended sediment transport in nonequilibrium situations. J. Hydr. Eng. Vol. 114, No. 10. p. 1157 - 1191.

CERC, 1977. Coastal Engineering Research Center. Shore Protection Manual, Vol. 1. Dept. of the Army. U.S. Gov. Printing Office. Washington, D.C. USA.

CERC, 1984. Coastal Engineering Research Center. Shore Protection Manual, Vol. 1. Dept. of the Army. U.S. Gov. Printing Office. Washington, D.C. USA.

Chao, S.Y., P.T. Shaw and J. Wang, 1995. Wind relaxation as a possible cause of the South China Sea Warm Current. J. Oceanography. Vol. 51, p. 111-132.

Chapman, D.C., 1985. Numerical treatment of cross-shelf open boundaries in a barotropic coastal ocean model. J. Phys. Ocean. 15, 1060-1075.

Chase, R. R. P., 1979. Settling behavior of natural aquatic particulates. Limnol. Oceanogr. Vol. 24, No. 3. p. 417 - 426.

Chen, M.P., 1993. The surface sediments of the South China Sea. Report presented at the First Working Group Meeting on Marine Scientific Research in the South China Sea. Annex N.

Chen, Z., 1992. Sediment concentration and sediment transport due to action of waves and a current. Communications on Hydraulic and Geotechnical Engineering. Rep. No. 92-9. Faculty of Civil Engineering. Delft University of Technology. Delft, The Netherlands.

Cleveland, J. S. and A. D. Weidemann, 1993. Quantifying absorption by aquatic particles: A multiple scattering correction for glass-fiber filters. Limnol. Oceanogr. Vol. 38, No. 6. p. 1321 - 1327.

Cole, P. and G. V. Miles, 1983. Two-dimensional model of mud transport. J. Hydr. Eng. Vol. 109. No. 1.

Crean, P. B., 1983. The development of rotating, non-linear numerical models simulating barotropic mixed tides in a complex coastal system located between vancouver Island

and the Mainland. Canadian Technical Report of Hydrography and Ocean Sciences. No. 31.

Davies, A.M. and J.E. Jones, 1992. A three dimensional model of the M_2, S_2, N_2, K_1 and O_1 tides in the Celtic and Irish Seas. Prog. Oceanog. Vol. 29, 197-234.

De las Alas, J. G., 1986. Prediction of the levels and trends of non-oil pollutants in the Lingayen Gulf. University of the Philippines Science Research Foundation. Diliman, Quezon City.

Deleersnijder, E., A. Norro and E. Wolanski, 1992. A three-dimensional model of the water circulation around an island in shallow water. Cont. Shelf Res. Vol. 12, No. 7/8, 891-906.

DePinto, J. V., W. Lick and J.F. Paul, 1994. Transport and transformation of contaminants near the sediment-water interface. Lewis Publishers.

De Vriend, H. J., 1991. Mathematical modelling and large-scale coastal behaviour. Part 2: Predictive models. J. Hydr. Res. Vol. 29, No. 6. p. 741 - 753.

Dyer, K.R., 1986. Coastal and Estuarine Sediment Dynamics. John Wiley & Sons. Chichester, England. pp. 342.

Effler, S. W., 1988. Secchi disc transparency and turbidity. J. Env. Eng. Vol. 114, No. 6. p. 1436 - 1447.

Ekebjaerg, L. and P. Justesen, 1991. An explicit scheme for advection-diffusion modelling in two dimensions. Comp. Meth. Appl. Mech. Eng. Vol. 88. p. 287 - 297.

EIA, 1994. Study of the Urgent Rehabilitation and Improvement works for the Agno River Flood Control Project. U.P. Science Research Found, Inc. Diliman, Q.C. Philippines.

Enggrob H. G. and K. W. Olesen, 1996. A hybrid sediment transport modelling system for rivers and estuaries. Arch. Hydrobiol. Spec. Issues. Advanc. Limnol. Vol. 47, p. 491 - 495.

Fenton, J. D. and W. D. McKee, 1990. On calculating the lengths of water waves. Coastal Engineering, Vol. 14, p. 499 - 513.

Flather, R. A. and N. S. Heaps, 1975. Tidal computations for Morecambe Bay. Geophys. J. R. Astr. Soc. p 489 - 517.

Galappati, G. and C. B. Vreugdenhil, 1986. A depth-integrated model for suspended sediment transport. J. Hydr. Res. Vol. 23 No. 4. p. 359 - 377.

Garver, S. A., D. A. Siegel and B. G. Mitchell, 1994. Variability in near-surface particulate absorption spectra: What can a satellite ocean color imager see? Limnol. Oceanogr. Vol. 39, No. 6. p. 1349 - 1367.

GESAMP (IMO/FAO/UNESCO/WMO/WHO/IAEA/UN/UNEP Joint Group of Experts on the Scientific Aspects of Marine Pollution), 1991. Coastal Modelling, GESAMP Reports and Studies No. 43, 192 pp.

Gill, A.E., 1982. Atmosphere-Ocean Dynamics. Academic Press, San Diego, California. 643 pp.

Gordon, H. R., 1989. Dependence of the diffuse reflectance of natural waters on the sun angle. Limnol. Oceanogr. Vol. 34, No. 8. p. 1484 - 1489.

Gordon, H. R., 1989. Can the Lambert-Beer law be applied to the diffuse attenuation coefficient of ocean water? Limnol. Oceanogr. Vol. 34, No. 8. p. 1389 - 1409.

Groen, P. and R. Dorresteyn, 1976. Zeegolven. In: Opstellen op oceanografisch en maritiem gebied. Vol. 11, KNMI, Staatsdrukkerij, Den Haag, 1-124.

Hearn, C.J. and I.N Parker, 1988. Hydrodynamic processes on the Ningaloo coral reef, Western Australia. Proc. 6th Intl. Coral Reef Symp., Australia. Vol 2., 407-502.

Holthuijsen, L.H., N. Booij, and T.H.C. Herbers, 1989. A prediction model for stationary, short-crested waves in shallow water with ambient currents. Coastal Engineering, 13, 23-54.

Hootsmans, M.J.M, and J.E. Vermaat, 1991. Macrophytes, a key to understanding changes caused by eutrophication in shallow freshwater ecosystems. IHE Report Series 21.

Iturriaga, R. and D. A. Siegel, 1989. Microphotometric characterization of phytoplankton and detrital absorption in the Sargasso Sea. Limnol. Oceanogr. Vol. 34, No. 8. p. 1706 - 1726.

Justesen, P., K.W. Olesen, and H.J. Vested. High-accuracy modelling of advection in two and three-dimensions. In: Proc. IAHR 23rd Congress (Ottawa, Canada, 1989) D-239, D-246.

Kirk, J. T. O., 1980. Spectral absorption properties of natural waters: Contribution of the soluble and particulate fractions to light absorption in some inland waters of South-eastern Australia. Aust. J. Mar. Freshwater Res. Vol 31. p. 287 - 296.

Kirk, J. T. O., 1981. Estimation of the scattering coefficient of natural waters using underwater irradiance measurements. Aust. J. Mar. Freshwater Res. Vol. 32, p. 533 - 539.

Kirk, J. T. O., 1977. Use of a quanta meter to measure attenuation and underwater reflectance of photosynthetically active radiation in some inland and coastal South-eastern Australian waters. Aust. J. Mar. Freshwater Res. Vol. 28, p. 9 -21.

Kirk, J. T. O., 1979. Spectral distribution of photosynthetically active radiation in some South-eastern Australian waters. Aust. J. Mar. Freshwater Res. Vol. 30, p. 81 - 91.

Kirk, J. T. O., 1976. Yellow substance (Gelbstoff) and its contribution to the attenuation of photosynthetically active radiation in some inland and coastal South-eastern Australian waters. Aust. J. Mar. Freshwater Res. Vol 27, p. 61 - 71.

Koutitas, C.G., 1988. Mathematical models in coastal engineering. Pentech Press, London. 156 pp.

Kristensen, P., M. Sondergaard and E. Jeppesen, 1992. Resuspension in a shallow eutrophic lake. Hydrobiologia 228: 101-109.

Lee, D. H., K. W. Bedford and C. C. J. Yen, 1994. Storm and entrainment effects on tributary sediment loads. J. Hydr. Eng. Vol. 120, No. 1. p. 81 - 103.

Leonards, B. P., 1991. The ULTIMATE conservative difference scheme applied to unsteady one-dimensional advection. Comp. Meth. Appl. Mech. Eng. Vol 88. No. 1. p. 17 - 74.

LeVeque, R. J., 1996. High-resolution conservative algorithms for advection in incompressible flow. SIAM J. Numer. Anal. Vol. 33, No. 2. p. 627 - 665.

Levitus, S., 1982. Climatological atlas of the world ocean. NOAA Professional paper No. 13, U. S. Government Printing Office, Washington D. C., 173 p.

Li, C. W. and T. S. Yu, 1994. Conservative characteristics-based schemes for mass transport. J. Hydr. Eng. Vol. 120. No. 9.

Lick, W., 1982. Entrainment, deposition, and transport of fine-grained sediments in lakes. Hydrobiologia 91, 31-40.

Lick, W., H. Huang and R. Jepsen, 1993. Flocculation of fine-grained sediments due to differential settling. J. Geophys. Res. Vol. 98, No. C6, p. 10,279 - 10,288.

Lijklema, L., R.H. Aalderink, G. Blom and E.H.S. Van Duin, 1991. Sediment transport in shallow lakes - two case studies related to eutrophication. In: Transport and transformation of contaminants near the sediment-water interface. DePinto, J.V. (ed). Springer Verlag, Berlin.

Luettich, R.A., 1987. Sediment resuspension in a shallow lake. Ph.D. Thesis. Massachusetts Inst. of Tech., Cambridge, Mass., U.S.A.: 311 pp.

Luettich, R.A., R.F. Harleman and L. Somlyody, 1990. Dynamic behavior of suspended sediment concentrations in a shallow lake perturbed by episodic wind events. Limnol. Oceanogr., 35(5): 1050-1067.

Maaliw, M.L.L., N. Bermas, R. Mercado and F. Guarin, 1989. Preliminary results of a water quality baseline study of Lingayen Gulf, p.83-91. In: G. Silvestre, E. Miclat and T.-E. Chua (eds). Towards sustainable development of the coastal resources of Lingayen Gulf, Philippines. ICLARM Conference Proceedings 17, 200 p. Philippine Council for Aquatic and Marine Research and Development, Los Baños, Laguna, and International Center for Living Aquatic Resources Management, Makati, Metro Manila, Philippines.

Maaliw, M.L.L., 1990. Pollution. In: The coastal environmental profile of Lingayen Gulf, Philippines. McManus, L.T. and C. Thia-Eng (eds.) ICLARM, 32-37.

McManus, J. W., C. L. Nanola, Jr., R. B. Reyes, Jr. and K. N. Kesner, 1992. Resource ecology of the Bolinao reef system. ICLARM Stud. Rev. 22, 117 p.

McManus, L. T. and T.-E. Chua, 1990. The coastal environmental profile of Lingayen Gulf, Philippines. ICLARM Technical Reports 22, 69.

Megard, R. O. and T. Berman, 1989. Effects of algae on the Secchi transparency of the southeastern Mediterranean Sea. Limnol. Oceanogr. Vol. 34, No. 8. p. 1640 - 1655.

Mehta, A.J., 1988. Laboratory studies on cohesive sediment deposition and erosion. In: Physical processes in estuaries. J. Dronkers and W. van Leussen, eds. Springer-Verlag, Berlin, 1988. 427-445.

Mehta, A.J., E. J. Hayter, R. Parker, R. B. Krone, and A. M. Teeter, 1989. Cohesive sediment transport. I: Process description. J. Hydr. Eng. Vol. 115, No. 8. p. 1076 - 1093.

Mehta, A. J., W. H. McAnally, Jr., E. J. Hayter, A. M. Teeter, D. Schoellhamer, S. B. Heltzel and W. P. Carey. Cohesive sediment transport. II: Application. J. Hydr. Eng. Vol. 115, No. 8. p. 1094 - 1110.

Mobley, C. D., 1989. A numerical model for the computation of radiance distributions in natural waters with wind-roughened surfaces. Limnol. Oceanogr. Vol. 34, No. 8. p. 1473 - 1483.

Morrow, J. H., W. S. Chamberlin and D. A. Kiefer, 1989. A two-component description of spectral absorption by marine particles. Limnol. Oceanogr. Vol. 34, No. 8. p. 1500 - 1509.

Murphy, P. J. and A. M. Aguirre, 1985. Bed load or suspended load. J. Hydr. Eng. Vol. 111, No. 1. p. 93 - 107.

Nakata, K., 1989. A simulation of the process of sedimentation of suspended solids in the Yoshii river estuary. Hydrobiologia 176/177: 431-438.

NWRC, 1976. Philippine water resources. National water resource commission. Quezon City, Philippines

Otsubo, K. and K. Muraoka, 1988. Critical shear stress of cohesive bottom sediments. J. Hydr. Eng. Vol. 114, No. 10. p. 1241 - 1256.

Parsons, T.R., Y. Maita and C. M. Lalli, 1984. A manual of chemical and biological methods for seawater analysis. p 101-109.

Parchure, T. M. and A. J. Mehta, 1985. Erosion of soft cohesive sediment deposits. J. Hydr. Eng. Vol. 111, No. 10.

Pejrup, M., 1988. Suspended sediment transport across a tidal flat. Marine Geology, 82: 187-198.

Pohlmann, T., 1987. A three-dimensional circulation model of the South China Sea. In: Three-dimensional models of marine and estuarine dynamics. p. 245-268. J. J. Nihoul and B. M. Jamart (eds). Elsevier, New York.

Preisendorfer, R.W., 1961. Application of radiative transfer theory to light measurements in the sea. Union Geod. Geophys. Inst. Monogr., 10.

Prieur, L. and S. Sathyendranath, 1981. An optical classification of coastal and oceanic waters based on the specific spectral absorption curves of phytoplankton pigments, dissolved organic matter, and other particulate materials. Limnol. Oceanogr. Vol. 26, No. 4. p. 671 - 689.

Roesler, C. S. and M. J. Perry, 1989. Modeling in situ phytoplankton absorption from total absorption spectra in productive inland marine waters. Limnol. Oceanogr. Vol. 34, No. 8. p. 1510 - 1523.

Sahin, V. and Hall, M. J., 1996. The effects of afforestation and deforestation on water yields. J. Hydrology. Vol. 178, p. 293 - 309.

Santos, G. Jr., A. F. Ramos, E. A. Marcelo, C. A. Petrache, L. G. Fernandez, M. K. Castillo and R. Almoneda, 1991. Notes on the geochemical distribution of elements in Lingayen Gulf sediments. Proc. 1st National Symposium in Marine Science.

Shaw, P.T. and S.Y. Chao, 1994. Surface circulation in the South China Sea. Deep Sea Research. Vol. 41, No. 11/12. p. 1663-1683.

Sheng, Y. P. and W. Lick, 1979. The transport and resuspension of sediments in a shallow lake. J. Geophys. Res., Vol. 84, No. C4. p. 1809 - 1826.

Somlyody, L. and L. Koncsos, 1991. Influence of sediment resuspension on the light conditions and algal growth in Lake Balaton. Ecological Modelling 57: 173-192.

Smith, I.R. Hydroclimate: The influence of water movement on freshwater ecology. Elsevier, London. 285 p.

Smith, R. C., B. B. Prezelin, R. R. Bidigare and K. S. Baker, 1989. Biooptical modeling of photosynthetic production in coastal waters. Limnol. Oceanogr. Vol. 34, No. 8. p. 1524 - 1544.

Spaulding, M.L. and C.H. Beauchamp, 1983. Modeling tidal circulation in coastal seas. J. Hydr. Eng. Vol. 109, No. 1. 116 - 132.

Stone, M., M.C. English and G. Mulamoottil, 1991. Sediment and nutrient transport dynamics in two tributaries of Lake Erie: A numerical model. Hydrological Processes, 5: 371-382.

Teeter, A. M., 1986. Vertical transport in fine-grained suspension and nearly-deposited sediment. Estuarine Cohesive Sediment Dynamics, Lecture Notes on Coastal and Estuarine Studies, 14. Springer Verlag, p. 126-149.

Teisson, C. 1992. Cohesive suspended sediment transport: feasibility and limitations of numerical modeling. J. Hydr. Res. Vol. 29, No. 6. p. 755 - 769.

Umita, T., T. Kusuda, Y. Awaya, M. Onuma and T. Futawatari, 1984. The behaviour of suspended sediments and muds in an estuary. Wat. Sci. Tech. 17: 915-927.

Van Den Berg, J. H. and A. Van Gelder, 1993. Prediction of suspended bed material transport in flows over silt and very fine sand. Water Resources Res. Vol. 29, No. 5. p. 1393 - 1404.

Van Duin, E.H.S., 1992. Sediment transport, light and algal growth in the Markermeer.

Ph.D. Thesis, Water Quality Section, Dept. of Nature Conservation, Wageningen Agricultural University, The Netherlands.

Van Leussen, W. 1988. Aggregation of particles, settling velocity of mud flocs: A review. In: J. Dronkers and W. van Leussen. Physical processes in estuaries. Springer-Verlag, Berlin: 348-551.

Van Leussen, W., 1994. Estuarine macroflocs and their role in fine-grained sediment transport. Ph.D. Thesis, Faculteit Aardwetenschappen, University of Utrecht, The Netherlands.

Van Rijn, L. C., 1982. Sediment Transport, Part II: Suspended Load Transport. J. Hydr. Eng. Vol. 110, No. 11. p. 1613 - 1639.

Van Rijn, L. C., 1986. Mathematical modelling of suspended sediment in non-uniform flows. J. Hydr. Eng. Vol. 112, No. 6. p. 433 - 455.

Van Rijn, L. C., 1993. Principles of sediment transport in rivers, estuaries and coastal seas. Aqua Publications Amsterdam, The Netherlands.

Villanoy, C. 1988. Inferred circulation of the Bolinao coral reef. Marine Science Institute Technical Report. University of the Philippines, Diliman, Quezon City.

Wang, J.D., A. F. Blumberg, H. L. Butler and P. Hamilton, 1990. Transport prediction in partially stratified tidal water. J. Hydr. Eng. Vol 116, No. 3. p 380 - 396.

Wang, Z. B., 1989. Mathematical modelling of morphological processes in estuaries. Comm. on Hydraulic and Geotechnical Eng'g., Rep. No. 89-1, Dept. of Civil Eng'g. Delft University of Technology. Delft, The Netherlands.

Wang, Z. B. and J. S. Ribberink, 1986. The validity of a depth-integrated model for suspended sediment transport. J. Hydr. Res. Vol 24, No. 1. p. 53 - 67.

Washburn, L., B. H. Jones, A. Bratkovich, T. D. Dickey and M.S. Chen. Mixing, dispersion, and resuspension in vicinity of ocean wastewater plume. J. Hydr. Eng. Vol. 118, No. 1. p. 38 - 58.

Weidemann, A. D. and T. T. Bannister, 1986. Absorption and scattering coefficients in

Irondequoit Bay. Limnol. Oceanogr. Vol. 31, No. 3. p. 567 - 583.

Wilcock, P. R., 1993. Critical shear stress of natural sediments. J. Hydr. Eng. Vol. 119, No. 4. p. 491 - 503.

Wu, J., 1982. Wind stress coefficient from breeze to hurricane. J. Geophys. Res. Vol. C5.

Wyrtki, K., 1961. The Naga Report. Scientific results of marine investigation of the South China Sea and the Gulf of Thailand. Vol. 2. Physical Oceanography of the Southeast Asian Waters. Univ. of California Press, 195 pp.

Xian-Xou, J., 1992. Quasi-three dimensional modelling of flows and application to a solute transport. Ph.D. Thesis. Dept. of Civil Engineering. Delft University of Technology, Delft, The Netherlands.

Yamamoto, S. and T. Chiba, 1994. Sedimentation of terrigenous red soils in coastal area adjacent to an estuary and coral reefs, the Okkubi River Estuary, Okinawa Island, Japan. J. Oceanogr. Vol. 50, p. 423 - 435.

Zaneveld, J. R. V., 1989. An asymptotic closure theory for irradiance in the sea and its inversion to obtain the inherent optical properties. Limnol. Oceanogr. Vol. 34, No. 8. p. 1442 - 1452.

Ziegler, K. and B. S. Nisbet, 1995. Long-term simulation of fine-grained sediment transport in Large Reservoir. J. Hydr. Eng. Vol. 121, No. 11.

Zoppou, C. and S. Roberts, 1993. Numerical solution of the advection-diffusion equation. In: Modelling change in environmental systems. A. J. Jakeman, M. B. Beck, and M. J. McAleer, eds., John Wiley and Sons, Chitchester, England. p. 77 -97.

Zyserman, J. A. and J. Fredsoe, 1994. Data analysis of bed concentration of suspended sediment. J. Hydr. Eng. Vol. 120, No. 9. p. 1021 - 1042.

Appendices

Appendix 1

Sediment Fractionation and Chlorophyll-a Determination

Bottom Sediment Fractionation

Fractionation of bottom sediments was undertaken in the laboratory using the following procedure. For each of the three fractions assumed, the corresponding sampling depth and sampling time are shown in Table (1).

1. Take 200 g of newly obtained wet sediment samples from the field. Put sample in a petri-dish. Mix sample, then measure the weight (sample + disc).

2. Take (with spoon) some sample for the experiments. Weigh dish again noting the difference in original weight (W_1).

3. Weigh an aluminum cup (W_c). Take a second sample from the petri-dish and measure the wet weight of cup and wet sample (W_2).

4. Dry the sample in the aluminum cup and measure the dry weight (W_3).

5. Fill a 2-liter cylinder with (filtered) seawater. Mix the sediment sample thoroughly in the cylinder and take suspensions of 500 ml each at the particular depths and time intervals given in Table (1). For the sand fraction, the time interval of 0 second means that a sample is taken while stirring (vigorously) the 2-l suspension.

6. Determine the dry weight of each of these 500 ml samples by the NEN (Dutch) procedure for TSS determination. The dry weight of F1, F2 and F3 are respectively W_{s1}, W_{s2} and W_{s3}.

Sample	Depth	Time Elapsed
F1 (Sand)	11 cm below water level	0 second
F2 (Silt)	22 cm below water level	11 seconds
F3 (Clay)	11 cm below water level	30 minutes

Table 1.   Sampling depths and time intervals for sediment fractionation.

7.   Estimate the relative contribution of sand (F1), silt (F2) and Clay (F3) to the bottom sediment using

$$\% \; F1 = [(W_{s1} - W_{s2}) \times 2 \times 100]/[0.5 \times W_1 \times (W_3 - W_c)/(W_2 - W_c)]$$

$$\% \; F2 = [(W_{s2} - W_{s3}) \times 2 \times 100]/[0.5 \times W_1 \times (W_3 - W_c)/(W_2 - W_c)]$$

$$\% \; F3 = [W_{s3} \times 2 \times 100]/[0.5 \times W_1 \times (W_3 - W_c)/(W_2 - W_c)]$$

Suspended Sediment Fractionation

From the sediment trap samples taken in the field, a 2-liter suspension is prepared in settling tubes (2-l graduated cylinders). Similar sampling depths and time intervals shown in Table (1) are used. The dry weights of the samples are also determined in accordance with the (Dutch) NEN methods.

Chlorophyll-a Determination

The procedure used for measuring Chlorophyll-a is outlined below. The method itself is based on the Manual of Chemical and Biological Methods for Seawater Analysis (Parsons et al. 1984) with slight modifications.

1.   Filter 2-liter seawater samples over a glass microfibre (GF/C) filters under 1/2 atmospheric pressure vacuum.

2.   Drain the filter thoroughly by suction and store or extract as necessary. When stored,

samples are wrapped completely with aluminum foils in petri-dishes and placed in the freezer. Temperature should be kept below 0°C as much as possible and storage time should not exceed 2 weeks.

3. Place the filter in a centrifuge tube; add 15 ml of 90 % acetone to volume and shake thoroughly. Allow to stand overnight in a dark place (preferably refrigerated).

4. Centrifuge the contents of each tube at room temperature for 10 minutes under 2000 rpm.

5. Decant the supernatant into a spectrophotometer cuvette and measure the absorbances at wavelengths of 750, 664, 647 and 630 nm without delay.

6. Correct each extraction for a small turbidity blank by subtracting the 750 nm from the 664, 647 and 630 nm absorptions.

7. Calculate the amount of pigment in the original seawater sample using $C = 11.85E_{664}-1.54E_{647}-0.08E_{630}$ where E is the corrected absorbance at different wavelengths.

8. The amount of Chlorophyll-a (μg/l) is then calculated from (C x v)/(V x P) where v is the volume of acetone in ml, V is the volume of seawater used and P is the path length characteristic of the spectrophotometer cuvette used.

Except for the addition of several drops of $MgCO_3$ solution to the samples during filtration, the above procedure is similar to the method outlined in Parsons et al. (1984). The limit of detection of chlorophyll-a with this method is about 0.03 μg/l which is generally acceptable.

Appendix 2

Numerical Solution of the 3rd Order Transport Equation

Using upstream differences for the spatial derivatives of Equation (5.49), the following finite difference form is obtained (Ekebjaerg and Justesen 1991);

$$
\begin{aligned}
c_{i,j}^{n+1} = \; & c_{i,j}^{n}[1 - 2(\tfrac{1}{2}C_x^2 + D_x) + C_x C_y - 2(\tfrac{1}{2}C_y^2 + D_y) \\
& - 3(\tfrac{1}{6}C_x - \tfrac{1}{6}C_x^3 - C_x D_x) - 3(\tfrac{1}{6}C_y - \tfrac{1}{6}C_y^3 - C_y D_y) \\
& - 2(\tfrac{1}{2}C_x C_y - \tfrac{1}{2}C_x^2 C_y - C_y D_x) - 2(\tfrac{1}{2}C_x C_y - \tfrac{1}{2}C_x C_y^2 - C_x D_y)] \\
& + c_{i+1,j}^{n}[-\tfrac{1}{2}C_x + (\tfrac{1}{2}C_x^2 + D_x) + (\tfrac{1}{6}C_x - \tfrac{1}{6}C_x^3 - C_x D_x) \\
& \qquad + (\tfrac{1}{2}C_x C_y - \tfrac{1}{2}C_x^2 C_y - C_y D_x)] \\
& + c_{i,j+1}^{n}[-\tfrac{1}{2}C_y + (\tfrac{1}{2}C_y^2 + D_y) + (\tfrac{1}{6}C_y - \tfrac{1}{6}C_y^3 - C_y D_y) \\
& \qquad + (\tfrac{1}{2}C_x C_y - \tfrac{1}{2}C_x C_y^2 - C_x D_y)] \\
& + c_{i-1,j}^{n}[\tfrac{1}{2}C_x + (\tfrac{1}{2}C_x^2 + D_x) - C_x C_y + 3(\tfrac{1}{6}C_x - \tfrac{1}{6}C_x^3 - C_x D_x) \\
& + (\tfrac{1}{2}C_x C_y - \tfrac{1}{2}C_x^2 C_y - C_y D_x) + 2(\tfrac{1}{2}C_x C_y - \tfrac{1}{2}C_x C_y^2 - C_x D_y)] \\
& + c_{i,j-1}^{n}[\tfrac{1}{2}C_y - C_x C_y + (\tfrac{1}{2}C_y^2 + D_y) + 3(\tfrac{1}{6}C_y - \tfrac{1}{6}C_y^3 - C_y D_y) \\
& + 2(\tfrac{1}{2}C_x C_y - \tfrac{1}{2}C_x^2 C_y - C_y D_x) + (\tfrac{1}{2}C_x C_y - \tfrac{1}{2}C_x C_y^2 - C_x D_y)] \\
& + c_{i-2,j}^{n}[-(\tfrac{1}{6}C_x - \tfrac{1}{6}C_x^3 - C_x D_x)] \\
& + c_{i,j-2}^{n}[-(\tfrac{1}{6}C_y - \tfrac{1}{6}C_y^3 - C_y D_y)] \\
& + c_{i-1,j-1}^{n}[C_x C_y - (\tfrac{1}{2}C_x C_y - \tfrac{1}{2}C_x^2 C_y - C_y D_x) \\
& \qquad - (\tfrac{1}{2}C_x C_y - \tfrac{1}{2}C_x C_y^2 - C_x D_y)] \\
& + c_{i-1,j+1}^{n}[-(\tfrac{1}{2}C_x C_y - \tfrac{1}{2}C_x C_y^2 - C_x D_y)] \\
& + c_{i+1,j-1}^{n}[-(\tfrac{1}{2}C_x^2 C_y - C_y D_x)]
\end{aligned} \tag{1}
$$

Here, C is the Courant number defined in the x and y-directions as

$$C_x = u\frac{\Delta t}{\Delta x} \quad , \quad C_y = v\frac{\Delta t}{\Delta y} \tag{2}$$

and D is a dimensionless diffusivity given by

$$D_x = K_x\frac{\Delta t}{\Delta x^2} \quad , \quad D_y = K_y\frac{\Delta t}{\Delta y^2} \tag{3}$$

The original upstream scheme proposed by Leonard solves Equation (1) by making use of interpolating polynomials in a transport function defined in one dimension. Ekebjaerg and Justesen (1991) extended Leonard's scheme by defining the transport functions in two-dimensions. In this study, the same two-dimensional transport equation is used with the addition of the resuspension and sedimentation fluxes as source and sink terms. The extended transport equation solved in the staggered grid used in the present application is given by

$$c_{i,j}^{n+1} = c_{i,j}^n + [(Tx_{i-1,j}^n - Tx_{i,j}^n) + (Ty_{i,j-1}^n - Ty_{i,j}^n)] + \Delta t\frac{(\phi_r - \phi_s)}{h} \tag{4}$$

where Tx and Ty are transport functions in the x and y-directions respectively. The basic grid element in Figure (1) shows the positions of the transport functions relative to the sediment concentration c.

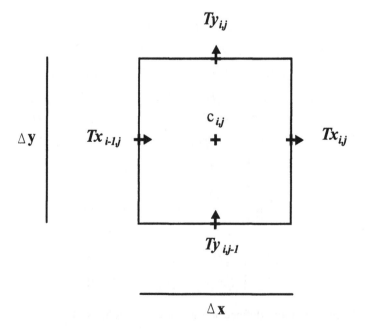

Figure 1. Basic grid element of the present transport model.

In accordance with Equation (1), the transport functions can be estimated from

$$Tx_{i,j}^n = A_1 c_{i+1,j}^n + A_2 c_{i,j}^n + A_3 c_{i-1,j}^n + A_4 c_{i,j+1}^n + A_5 c_{i,j-1}^n - D_x c_{i+1,j}^n + D_x c_{i,j}^n$$
$$Ty_{i,j}^n = B_1 c_{i,j+1}^n + B_2 c_{i,j}^n + B_3 c_{i,j-1}^n + B_4 c_{i+1,j}^n + B_5 c_{i-1,j}^n - D_y c_{i,j+1}^n + D_y c_{i,j}^n$$

(5)

The interpolating weights A and B can be obtained from

$$A_1 = (\tfrac{1}{6}C_x^2 - \tfrac{1}{2}C_x + \tfrac{1}{3} + D_x)C_x$$
$$A_2 = (-\tfrac{1}{3}C_x^2 + \tfrac{1}{2}C_x + \tfrac{5}{6} - \tfrac{1}{2}C_x C_y - \tfrac{1}{2}C_y^2 + \tfrac{1}{2}C_y - 2D_x - 2D_y)C_x$$
$$A_3 = (-\tfrac{1}{6} + \tfrac{1}{6}C_x^2 + D_x)C_x$$
$$A_4 = (-\tfrac{1}{2}C_y + \tfrac{1}{2}C_y^2 + D_y)C_x$$
$$A_5 = (\tfrac{1}{2}C_x C_y + D_y)C_x$$

(6)

$$B_1 = (\tfrac{1}{6}C_y^2 - \tfrac{1}{2}C_y + \tfrac{1}{3} + D_y)C_y$$
$$B_2 = (-\tfrac{1}{3}C_y^2 + \tfrac{1}{2}C_y + \tfrac{5}{6} - \tfrac{1}{2}C_y C_x - \tfrac{1}{2}C_x^2 + \tfrac{1}{2}C_x - 2D_y - 2D_x)C_y$$
$$B_3 = (-\tfrac{1}{6} + \tfrac{1}{6}C_y^2 + D_y)C_y$$
$$B_4 = (-\tfrac{1}{2}C_x + \tfrac{1}{2}C_x^2 + D_x)C_y$$
$$B_5 = (\tfrac{1}{2}C_y C_x + D_x)C_y$$

(7)

It should be noted that the transport functions Tx and Ty are defined at the same locations as u and v in the hydrodynamic model, i.e. at the edges of a grid cell as shown in Figure (1). This gives vanishing sediment transport normal to solid boundaries such as islands and coasts, as normal flow velocities are zero in the present grid setup. This is one of the obvious advantages of the present transport formulation (Ekebjaerg and Justesen 1991).

The upstream nature of the scheme implies that the spatial indices i and j have to be changed in accordance with the local hydrodynamic conditions. The locations of the interpolating weights are thus adjusted when the flow velocities change directions. Figure (2) shows the positioning of the variables in cases when *a)* $u > 0$, $v > 0$, and *b)* $u < 0$, $v > 0$. Note that the concentrations c's are positioned along with the interpolating polynomials, i.e. at the center of the grid cell.

The resuspension and sedimentation fluxes are likewise estimated at the center of the grid cells similar to the sediment concentrations. While no special numerical techniques are required for their solutions, a correct balance of these two fluxes should nevertheless be ensured in order to obtain positive concentrations. The addition of the resuspension and sedimentation processes makes the present scheme suitable for suspended sediment transport modelling in coastal areas in the presence of both currents and waves.

For numerical stability, the general criterion imposed on the present scheme involves the inequality

$$(|C_x| + |C_y| + |D_x| + |D_y|) \leq 1 \qquad (8)$$

In a series of experiments, it was found by Ekebjaerg and Justesen (1991) that the scheme is stable if the convective Courant numbers do not exceed unity. The inclusion of the diffusive terms was found to increase the stability of the numerical scheme as long as they do not dominate over the convective terms. If they dominate, the stability decreases. With flow velocities of less than 1 m s^{-1} and dispersion coefficients (K_x and K_y) of 10 m^2 s^{-1}, a time interval Δt of 60 seconds gives stable solutions in the present numerical scheme.

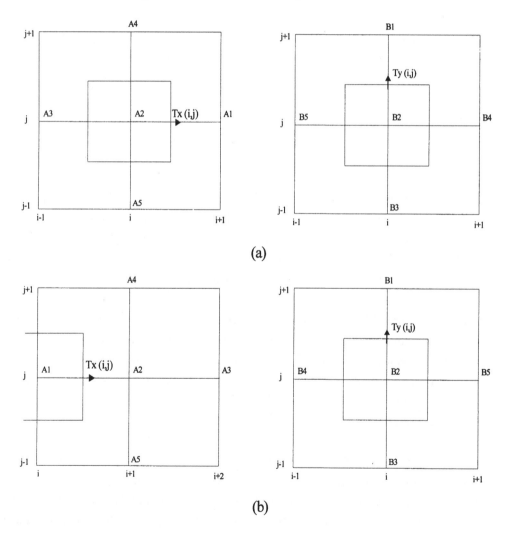

Figure 2. Determination of the transport functions for *a*) *u*, *v* > 0 and *b*) *u* < 0, *v* > 0 (Ekebjaerg and Justesen 1991).

Appendix 3

List of Symbols

The values of some parameters as used in the Cape Bolinao and Lingayen Gulf models are given in the following list. Parameter values, determined from calibration experiments, are given in the text.

a	reference height (level)	m
a	stress-related variable	m s^{-1}
a	absorption coefficient	m^{-1}
a_x	x-component of the stress variable	m s^{-1}
a_y	y-component of the stress variable	m s^{-1}
a_C	absorption coefficient of phytoplankton	m^{-1}
a_d	diffuse absorption coefficient	m^{-1}
a_o	average tidal amplitude	m
$a_{1,2,3,4}$	tidal amplitudes (O$_1$, K$_1$, M$_2$ and S$_2$ constituents)	m
A	absorption cross-section of phytoplankton	μm^2
A_h	diffusion coefficient (10)	m^2 s^{-1}
b	scattering coefficient	m^{-1}
b_{bd}	diffuse backscattering coefficient for downward irradiance	m^{-1}
b_{bu}	diffuse backscattering coefficient for upward irradiance	m^{-1}
c	phase speed of a boundary variable	m s^{-1}
c	depth-averaged sediment concentration	kg m^{-3}
c_a	reference concentration	kg m^{-3}
c_b	near-bed concentration	m^3 m^{-3}
c_B	sediment concentration at the boundary	kg m^{-3}
c_d	surface stress drag coefficient	-

c_m	maximum volumetric concentration	m^3 m^{-3}
c_o	background concentration	kg m^{-3}
C	Chezy coefficient	$m^{1/2}$ s^{-1}
C	grain-related Chezy coefficient	$m^{1/2}$ s^{-1}
C_L	non-dimensional variable in the Orlanski Radiation Condition	-
d	particle diameter	m
d_s	diameter of a spherical particle	m
d_{50}	mean particle diameter	m
D_*	dimensionless particle parameter	-
E	sediment pick-up rate	kg m^{-2} s^{-1}
E_d	downward irradiance	μE m^{-2} s^{-1}
E_u	upward irradiance	μE m^{-2} s^{-1}
E_z	downward irradiance at depth z	μE m^{-2} s^{-1}
E_o	incident downward irradiance	μE m^{-2} s^{-1}
F	Fetch (distance)	m
f	Coriolis parameter	s^{-1}
f_c	bed friction coefficient for current	-
f_w	bed friction coefficient for waves	-
g	gravitational acceleration (9.81)	m s^{-2}
G	absorption coefficient of gilvin measured at 380 nm	m^{-1}
h	total water depth	m
h_o	still water level	m
H_{cr}	critical wave height	m
H_{ref}	reference wave height	m
H_s	significant wave height	m
i	spatial index in the x-direction	-
j	spatial index in the y-direction	-
k	bottom friction coefficient (0.01)	-
k	constant in Equations (5.26 and 5.28)	-
k_d	extinction coefficient	m^{-1}
$k_{d,w}$	partial extinction coefficient due to pure water	m^{-1}
$k_{d,p}$	partial extinction coefficient due to suspended particulate	m^{-1}
$k_{d,c}$	partial extinction coefficient due to phytoplankton chlorophyll	m^{-1}
$k_{d,g}$	partial extinction coefficient due to gilvin	m^{-1}
k_r	resuspension constant (also floc erosion rate)	kg m^{-2} s^{-1}
$k_{s,c}$	current-related bed roughness height (0.1)	m
$k_{s,w}$	wave-related bed roughness height (0.01)	m
K	constant in Equations (5.29 and 5.43)	-

$K_{x,y,z}$	dispersion coefficients (10 for x and y-axes only)	$m^2\ s^{-1}$
L_s	significant wave length	m
m	empirical constant in Equation (5.43)	-
M	Partheniades erosion parameter	$kg\ m^{-2}\ s^{-1}$
n	wave frequency	s^{-1}
n	time level	-
N	number of algal cells	-
p	probability of deposition	-
$p_{1,2,3,4}$	tidal phases (O_1, K_1, M_2 and S_2)	rad
P_e	Peclet number	-
q_b	bed load transport rate	$kg\ m^{-1}\ s^{-1}$
q_s	suspended load transport rate	$kg\ m^{-1}\ s^{-1}$
R	irradiance reflectance	-
s	smoothing coefficient for bathymetry (0.5)	-
s	specific weight of sediment (ρ_s/ρ)	-
S	practical salinity scale	
t_d	time scale for dispersion	s
t_s	time scale for settling	s
T_s	significant wave period	s
T	dimensionless bed-shear stress parameter	-
Tx	transport function in the x-direction	$kg\ m^{-3}$
Ty	transport function in the y-direction	$kg\ m^{-3}$
u	x-component of depth-averaged flow velocity	$m\ s^{-1}$
u_b	sediment particle velocity	$m\ s^{-1}$
u_{cr}	critical depth-averaged flow velocity	$m\ s^{-1}$
u_m	maximum wave orbital velocity	$m\ s^{-1}$
u_{mc}	critical wave orbital velocity	$m\ s^{-1}$
u_{surf}	x-component of the surface current	$m\ s^{-1}$
v_{surf}	y-component of the surface current	$m\ s^{-1}$
u_*	bed shear velocity	$m\ s^{-1}$
v	y-component of current velocity	$m\ s^{-1}$
V	magnitude of flow velocity	$m\ s^{-1}$
w_s	sediment settling velocity	$m\ s^{-1}$
$w_{s,m}$	particle settling velocity in a mixture	$m\ s^{-1}$
W	wind speed	$m\ s^{-1}$
W_x	wind speed component in the x-direction	$m\ s^{-1}$
W_y	wind speed component in the y-direction	$m\ s^{-1}$
x,y,z	Cartesian coordinates	

z_m	mid-point of the euphotic zone	m
z_*	Rouse or suspension number	-
α	parameter in Equation (5.31) (0.5)	$m\ N^{-1/2}$
β	proportionality constant between sediment and fluid diffusion	-
δ_b	bed-load layer thickness	m
Δx	grid spacing in the x-direction ($= \Delta s$)	m
Δy	grid spacing in the y-direction ($= \Delta s$)	m
Δs	grid spacing (fine and coarse model = 500 m and 1000 m)	m
Δt	time interval (fine and coarse model = 10 s)	s
ϵ_f	fluid mixing coefficient	$m^2\ s^{-1}$
ζ	sea surface elevation	m
$\zeta(t)$	tidal forcing function	m
η	coefficient determining inclusion or exclusion of advection	-
θ	mobility parameter	-
θ_{cr}	critical mobility parameter	-
Θ	coefficient of form resistance	-
κ	Von Karman constant (0.4)	-
κ_p	specific extinction coefficient for suspended particulate	$m^2\ g^{-1}$
κ_c	specific extinction coefficient for algae	$m^2\ mg^{-1}$
κ_g	specific extinction coefficient for gilvin	-
$\kappa_{1,2,3}$	specific extinction coefficients of sediment fractions	$m^2\ g^{-1}$
λ	wavelength	nm
λ	non-dimensional constant in Equation (4.21)	-
λ_b	saltation length	m
μ	non-dimensional phase speed	-
$\mu(z)$	depth-dependent average cosine of a quanta of radiation	-
ν	kinematic viscosity coefficient of water	$m^2\ s^{-1}$
ν_h	diffusion coefficient	$m^2\ s^{-1}$
ν	eddy viscosity	$m^2\ s^{-1}$
ξ	dimensionless parameter in Equation (5.20)	-
ρ	water density (1025)	$kg\ m^{-3}$
ρ_a	air density (1.22)	$kg\ m^{-3}$
ρ_s	sediment density	$kg\ m^{-3}$
τ_{sx}	surface stress component in the x-direction	$N\ m^{-2}$
τ_{sy}	surface stress component in the y-direction	$N\ m^{-2}$
τ_{bx}	bottom stress component in the x-direction	$N\ m^{-2}$

τ_{by}	bottom stress component in the y-direction	N m^{-2}
τ_b	bed shear stress	N m^{-2}
τ_c	current-related shear stress	N m^{-2}
τ_{cr}	critical bed shear stress	N m^{-2}
τ_{cw}	current-wave shear stress	N m^{-2}
τ_w	wave-related shear stress	N m^{-2}
ϕ_r	resuspension flux	kg m^{-2} s^{-1}
ϕ_s	sedimentation flux	kg m^{-2} s^{-1}
ϕ	boundary variable denoting velocity or sea surface elevation	m s^{-1} or m
Φ	Latitude (16.33)	°
χ	dimensional coefficient in Equation (5.1)	s kg^{-1}
$\omega_{1,2,3,4}$	tidal frequencies (O_1, K_1, M_2 and S_2)	rad hr^{-1}
Ω	Earth's angular rotation rate (7.292 x 10^{-5})	rad s^{-1}
∇^2_h	2-D Horizontal Laplacian operator	

Curriculum Vitae

Paul C. Rivera was born in Isabela, Philippines on 29 November 1966. He finished his basic secondary education in a catholic school run by the La Salette priests in Cabatuan, Isabela. In 1988, he obtained his Bachelor of Science (Physics) degree in the same province and a year later, he moved to Manila to pursue graduate education. With a fellowship granted by the Philippine Atmospheric, Geophysical and Astronomical Services Administration (PAGASA), he finished his Master of Science (Meteorology) degree at the University of the Philippines, Quezon City in 1990 specializing in the numerical modelling of storm surges. While doing graduate courses in the Department of Meteorology and Oceanography, he worked as a University Research Associate from 1989 to 1991. In 1992, he was granted a fellowship to pursue a higher education under the auspices of the Wageningen Agricultural University (WAU) and the International Institute for Hydraulic and Environmental Engineering (IHE) in the Netherlands. In the same year, he obtained his Master of Science (Environmental Science and Technology) degree from both institutions. There, he studied the rudiments of water quality management with emphasis on field observational and modelling studies of sediment resuspension, sedimentation and light attenuation in a freshwater lake in the Netherlands. In 1993 and under the sponsorship of the Cooperation in Environmental Ecotechnology with Developing Countries (CEEDC) Project initiated by IHE and WAU, he started his PhD research studies on the hydrodynamics, sediment transport and light extinction of the marine environment in Cape Bolinao, Philippines.

Samenvatting en Conclusies

In deze studie is onderzoek uitgevoerd naar de waterbeweging, het sedimenttransport en de uitdoving van licht onder water in Cape Bolinao. Het studiegebied is gesitueerd in de monding van de Lingayen Gulf, aan de noord-west kust van de Filipijnen. Zoals in veel koraalrif ecosystemen is de lagune rijk aan flora en fauna soorten, zoals zeegrassen, algensoorten en koralen. Aanslibbing en eutrofiëring bedreigen de bio-diversiteit van die marine ecosysteem. Om deze redenen is een intensief onderzoek uitgevoerd naar de fysische milieufactoren, met de volgende doelstellingen:

1. Het geven van een kwantitatieve beschrijving van de circulatiepatronen, de sediment transport processen en de licht uitdoving in het zeewater in Cape Bolinao.

2. Het ontwikkelen van een geïntegreerd numeriek model voor de waterbeweging, het sediment transport en de licht uitdoving in Cape Bolinao en de omgeving daarvan, ten behoeve van milieu-effect-studies.

Het meetprogramma, dat bestond uit veldmetingen en laboratoriumexperimenten, werd uitgevoerd in een periode van bijna twee jaar, van augustus 1993 tot juni 1995. Er zijn verschillende meetlokaties geselecteerd rond Cape Bolinao, samen representatief voor het grootste deel van de met zeegrassen bezette gebieden. De toegepaste meetfrequenties waren gerelateerd aan de tijdschalen van de relevante processen en variabelen. Het meetprogramma omvatte metingen die met een hoge frequentie, één maal per uur, werden uitgevoerd, wekelijkse metingen en metingen die eens per kwartaal plaatsvonden. De stroomsnelheid en -richting, zwevende stofconcentratie (TSS), neerwaartse instraling en waterstand werden met een hoge frequentie gemeten op, afwisselend, vier lokaties vanaf een meetplatform. Behalve voor de zwevende stofconcentratie, die vanwege de beperkingen van het instrumentarium gemeten werd met een interval van 4 uur, werden deze metingen met een interval van 5 minuten gemeten, waarna de waarden per uur gemiddeld werden. Eens per 6 weken werd het platform en het daarop aanwezige instrumentarium verplaatst naar één van de 4 lokaties. Dit maakte het mogelijk om op alle gekozen lokaties metingen te verrichten in zowel het droge (Noordoost moesson) als in het regenachtige (Zuidwest moesson) seizoen op alle 4 plaatsen. De wekelijkse metingen werden uitgevoerd op 13 lokaties in het studiegebied en omvatten de oppervlakte-stromingen, de temperatuur, de saliniteit, de zwevende stofconcentratie, het as-vrij drooggewicht, de absorptie door humuszuren, de algenconcentratie en de neerwaartse straling. Hiermee werden gegevens verkregen voor de analyse van de maandelijkse en seizoensvariatie van de relevante variabelen. Eens per kwartaal werd de waterbodem bemonsterd en geanalyseerd. Gemeten werden de deeltjesgrootteverdeling en het organische stofgehalte. Eens per week werden sedimentmonsters verzameld met behulp van sedimentvallen. Met deze monsters en ook in de eens per kwartaal verzamelde bodemmonsters werden fractionerings-experimenten uitgevoerd om de deeltjesgrootteverdeling en de licht extinktie karakteristieken van een mengsel van sedimentdeeltjes te bepalen. Tevens werden in dit onderzoek de relevante meteorologische gegevens verzameld zoals de windsnelheid en richting en de regenval.

De resultaten van het veldonderzoek naar de waterbeweging in het studiegebied laten zien dat de stroming bepaald wordt door een niet-lineaire interactie tussen wind- en getijde-invloeden. Onder normale omstandigheden, zonder stormen, waren de oppervlakte stroomsnelheden binnen het rif laag ($\approx$ 10 cm s^{-1}). De door getijde beweging geïnduceerde stroomsnelheden waren van dezelfde orde van grootte, met uitzondering van die in enkele openingen in het rif en in het kanaal tussen Santiago Island en het vaste land, waar de stroomsnelheden vaak groter waren dan 10 cm s^{-1}. Waterstandsmetingen lieten een gemengd getijde zien, waarin een dagelijkse component domineert. Eens per veertien dagen was een semi-dagelijkse component in het getij duidelijk waarneembaar. De waterstandsvariatie was ongeveer 1 meter, hoewel deze vaak overschreden wordt gedurende springvloed. De saliniteit binnen het rif kende een duidelijke seizoensvariatie. De laagste saliniteit (S < 30) werd gemeten in het regenseizoen (Zuidwest moesson) als gevolg van toestroming van zoet water. De maximale saliniteitswaarden werden gemeten in het droge seizoen (Noordoost moesson), ten gevolge van de grotere verdamping en de verwaarloosbare zoetwatertoevoer in deze periode. De temperatuur in het ondiepe water van de lagune wordt gekenmerkt door een sterke dagelijkse en seizoensvariatie; dit ten gevolge van de variatie in de instraling. De gemeten zwevende stofconcentratie in het gebied rond Cape Bolinao was laag ($\approx$ 15 g m^{-3}). De variatie in de zwevende stofconcentratie werd onder normale omstandigheden, bij lage windsnelheden, vooral bepaald door getijdestroming. In stormachtige situaties werd het effect van wind op de variatie in de zwevende stofconcentratie significant. De sedimentatieflux in het kanaal tussen Santiago Island en het vaste land was gemiddeld 3.2 g m^{-2} hr^{-1}. Kleinere fluxen, gemiddeld 0.7 g m^{-2} hr^{-1}, werden gemeten binnen het rif. De gemeten algenconcentratie binnen het rif was erg laag ($\approx$ 0.45 mg m^{-3} Chlorofyl-a). Hogere gehalten, vaak meer dan 1 mg m^{-3}, werden in het kanaal waargenomen. De absorptie door humuszuren (bij 380 nm) was laag; gemiddeld 0.95 m^{-1}. Het gehalte organisch materiaal in de zwevende stof was gemiddeld 30%, gedurende de gehele meetperiode. Binnen het rif domineerde grof materiaal de bodemsamenstelling, terwijl in het kanaal meer fijn materiaal voorkwam. Het organische stofgehalte in de bodem was over het algemeen lager binnen het rif (< 5%) dan in het kanaal (over het algemeen 5 - 20%). De extinktie coëfficiënten in het studiegebied waren laag. Gemiddeld was de k_d 0.40 m^{-1}. Deze lage extinktiecoëfficiënt is het gevolg van de lage zwevende stof-, algen- en humuszurenconcentraties in het gebied.

De modelontwikkeling in deze studie was gericht op het ontwikkelen van numerieke modellen voor de analyse en voorspelling van waterbeweging, zwevende stoftransport en lichtuitdoving. Er werden twee onafhankelijke modellen ontwikkeld, te weten een model met een fijne resolutie voor het studiegebied zelf (het gebied rond Cape Bolinao) en een model met een grote plaatsstap voor de gehele Lingayen Gulf. Het model met de fijne resolutie werd ontwikkeld om de lokale processen te beschrijven en voorspellen, terwijl het grote model ontwikkeld werd om transportprocessen in de Lingayen Gulf te analyseren. Het waterbewegingsmodel dat toegepast werd in beide gebieden is een quasi drie-dimensionaal model in de zin dat stroomsnelheden op elke diepte kunnen worden geschat. In het model zijn realistische gegevens voor de drijvende krachten, wind en getijde, gebruikt. De getijdebeweging bestaat uit vier getijdecomponenten, die verantwoordelijk zijn voor de variatie in de waterstand in het gebied. Deze omvatten O_1, K_1, M_2 en S_2 getijden. Op basis van dit model voor wind- en getijde-geïnduceerde stroming, in combinatie met een model voor wind-geïnduceerde golven, werd een model voor het zwevende stoftransport ontwikkeld. Dit zwevende stof model heeft een derde-orde nauwkeurigheid in ruimte en tijd. Door de toevoeging van resuspensie en sedimentatie, als bron- en puttermen, beschrijft dit model de advectie en diffusie van gesuspendeerd sediment in Cape Bolinao en de Lingayen Gulf. Evenals in het waterbewegingsmodel is ook de numerieke oplossing die in het zwevende stofmodel gebruikt wordt gebaseerd op een eindige-differentie methode. Het derde orde schema is geschikt voor situaties met scherpe zwevende stofconcentratiegradiënten en kent weinig numerieke oscillatie zolang het waterbewegingsmodel numeriek stabiel is. Verder is de gebruikte oplossing massabehoudend, wat wenselijk is wanneer transportprocessen onderzocht worden. De met dit model gesimuleerde zwevende stofconcentratie werden, samen met gemeten bijdragen van algen, humuszuren en water zelf, gebruikt in een model voor de lichtuitdoving. Het lichtmodel is gebaseerd op de veronderstelling dat de totale extinktiecoëfficiënt een lineaire functie is van de bijdragen van de verschillende optisch actieve componenten. Gemiddelde algenconcentraties en absorptiecoëfficiënten veroorzaakt door humuszuren, gemeten in het veld en het laboratorium, werden in de simulatieberekeningen gebruikt. Op basis van deze lineaire benadering werd aangetoond dat algen en humuszuren in gelijke mate (elk ≈ 20 %) bijdrage en de lichtuitdoving rond Cape Bolinao. Anorganische en organische zwevende stof (exclusief de bijdrage van algen daaraan) dragen 52 % bij, terwijl water zelf ongeveer 8% van de lichtuitdoving veroorzaakt.

Met de numerieke modellen werden gevoeligheidsanalyses en modelcalibraties uitgevoerd. De parameterwaarden werden door middel van handmatige calibratie vastgesteld. Er werd een redelijke overeenkomst tussen gemeten en gesimuleerde waarden bereikt, met name voor de stroomsnelheden, waterstanden, totale zwevende stofconcentratie en extinktiecoëfficiënten.

Simulatieberekeningen met het geïntegreerde model lieten zien dat de invloed van rivieren in het studiegebied op de zwevende stofconcentratie en extintiecoëfficiënt binnen het rif klein is. Sediment vrachten in de nabijheid van de kaap blijken door het kanaal weg van het rif getransporteerd te worden. Verder bleken rivier afvoeren in het zuiden van de Lingayen Gulf weinig invloed op het koraalrif bij Bolinao te hebben. Model simulaties lieten zien dat 90% van de zwevende stofconcentratie binnen het rif toegeschreven kan worden aan interne, lokale resuspensie van bodemmateriaal, terwijl minder dan 10% afkomstig is van externe (rivier) bronnen. Het beperkte effect van de belasting vanuit rivieren wordt vooral veroorzaakt door het algemene circulatiepatroon in de Lingayen Gulf en de marine omgeving van Cape Bolinao.

DELFT

The aim of the International Institute for Infrastructural, Hydraulic and Environmental Engineering,
IHE Delft, is the development and transfer of
scientific knowledge and technological know-how
in the fields of transport, water and the environment.

Therefore, IHE organizes regular 12 and 18 month
postgraduate courses which lead to a Masters Degree.
IHE also has a PhD-programme based on research,
which can be executed partly in the home country.
Moreover, IHE organizes short tailor-made and
regular non-degree courses in The Netherlands as
well as abroad, and takes part in projects in various
countries to develop local educational training and
research facilities.

International Institute for
Infrastructural, Hydraulic and
Environmental Engineering

P.O. Box 3015
2601 DA Delft
The Netherlands

Tel.:   +31 15 2151715
Fax: +31 15 2122921
E-mail: ihe@ihe.nl
Internet: http://www.ihe.nl

T - #0650 - 101024 - C0 - 254/178/14 - PB - 9789054104087 - Gloss Lamination